Sustainable Textiles: Production, Processing, Manufacturing & Chemistry

Series Editor

Subramanian Senthilkannan Muthu, Chief Sustainability Officer
Green Story Inc.
Tung Chung, Hong Kong

This series aims to address all issues related to sustainability through the lifecycles of textiles from manufacturing to consumer behavior through sustainable disposal. Potential topics include but are not limited to: Environmental Footprints of Textile manufacturing; Environmental Life Cycle Assessment of Textile production; Environmental impact models of Textiles and Clothing Supply Chain; Clothing Supply Chain Sustainability; Carbon, energy and water footprints of textile products and in the clothing manufacturing chain; Functional life and reusability of textile products; Biodegradable textile products and the assessment of biodegradability; Waste management in textile industry; Pollution abatement in textile sector; Recycled textile materials and the evaluation of recycling; Consumer behavior in Sustainable Textiles; Eco-design in Clothing & Apparels; Sustainable polymers & fibers in Textiles; Sustainable waste water treatments in Textile manufacturing; Sustainable Textile Chemicals in Textile manufacturing. Innovative fibres, processes, methods and technologies for Sustainable textiles; Development of sustainable, eco-friendly textile products and processes; Environmental standards for textile industry; Modelling of environmental impacts of textile products; Green Chemistry, clean technology and their applications to textiles and clothing sector; Eco-production of Apparels, Energy and Water Efficient textiles. Sustainable Smart textiles & polymers, Sustainable Nano fibers and Textiles; Sustainable Innovations in Textile Chemistry & Manufacturing; Circular Economy, Advances in Sustainable Textiles Manufacturing; Sustainable Luxury & Craftsmanship; Zero Waste Textiles.

Subramanian Senthilkannan Muthu
Mukta Ramchandani

Vegan Alternatives for Leather

Springer

Subramanian Senthilkannan Muthu
Green Story, Canada
Tung Chung, Hong Kong

Mukta Ramchandani
Moraltive, Olten, Switzerland

Skema Business School
Sophia-Antipolis, France

ISSN 2662-7108 ISSN 2662-7116 (electronic)
Sustainable Textiles: Production, Processing, Manufacturing & Chemistry
ISBN 978-3-031-65364-3 ISBN 978-3-031-65365-0 (eBook)
https://doi.org/10.1007/978-3-031-65365-0

This Springer imprint is published by the registered company Springer Nature Switzerland AG
The registered company address is: Gewerbestrasse 11, 6330 Cham, Switzerland

Preface

Research in the area of vegan alternatives for leather is relatively new. Many academicians and industry experts are exploring this topic from their own fields, respectively. As a research topic develops, primary focus is on assessing the problem, finding a solution, and suggesting future directions. For the topic of leather, we know that environmental sustainability is of a huge discussion, and newer alternatives, especially non-animal sourced (vegan), are the need for the industry and consumers. While these newer alternatives are being developed and getting popular, their assessment for long-term usage is still limited. This is why our aim in this book is to provide a deeper understanding of these vegan alternatives and different perspectives are applied for this investigation. Ranging from the industry: especially luxury and fashion, producers and business in the field of material innovation, environmental sustainability; to academic: consumer behavior and marketing.

We hope that this book helps the readers in understanding this topic and further advancements in sustainability of the traditional animal based and vegan leather industry.

Tung Chung, Hong Kong — Subramanian Senthilkannan Muthu
Olten, Switzerland — Mukta Ramchandani

Acknowledgment

The idea of writing this book came from Dr. Subramanian Senthilkannan Muthu. A sincere gratitude, as without his expertise and our mutual interest in sustainability, this book would have not been possible. His knowledge truly inspires and paves the way for numerous researchers in this field, like me.

A heartfelt thanks to Dr. Ivan Coste-Maniere, for mentoring me on the track of sustainable luxury and fashion in France and introducing me to Dr. Subramanian.

I would also like to thank Adrián López Velarde the co-founder of Desserto from Mexico, Kim Hyunjoo the designer and artist at KHJ Studios South Korea, and my friend Min Zhou Carpentier the owner of EMZI Shoes Paris. Without their interviews, my research for this book would have been incomplete.

A lot goes behind the scenes, when it comes to the process of research and writing. And not often it is visible to the end readers. My family and especially my husband Benjamin Leitschuh have been very supportive in this process. I am grateful for their love, and thank them for having the patience toward my ubiquitous research and writing rhythms.

Finally, infinite gratitude toward my Gurus and God, for everything I learn is dedicated to them.

Olten, Switzerland — Mukta Ramchandani

Contents

Chapter 1
Definitions, Government Regulations, and Norms Within the "Vegan/Vegetarian Leather" Industry

1 Introduction

A long history of traditional animal-based leather exists for different products and usages across the world. With the influx of hazardous chemicals discharged into the environment, the leather industry is strongly criticized for its environmental impact. Moreover, animal cruelty found in exotic leather trade, such as leather derived from crocodiles, snakes, ostriches, etc., is a topic of concern for many consumers. From the perspective of sustainability, it is expected that companies adhere to sustainable production standards in terms of fair production with the use of proper tanning agents/materials, sustainable processes for waste discharge and effluent treatments, and minimizing the environmental impact of leather. Manufacturers, on the other hand, are limitedly embracing sustainable innovation within the leather industry and are confounded by the traditional methods of leather production and profitability. This has also led to the development of a parallel "vegan/vegetarian leather" industry, where several companies are increasingly producing products as an alternative to traditional leather. An important question arising here is how well these vegan leather alternatives perform vis-à-vis traditional leather. In this book, we discuss and assess how, beyond functionality, durability, and other sustainable paradigms, one can ascertain the long-term usability of such products and their market demand.

According to PETA (2022), "vegan leather" is described as all imitation leather which does not contain any animal skin/hide as raw material or animal-based adhesives and is visually like animal-based leather. A generic understanding of "vegan leather" is gaining traction among consumers in developed countries, evidenced by some luxury and fashion brands creating vegan leather collections, e.g., Stella McCartney using lab-grown mushroom leather called Mylo™ (Stella McCartney, 2021). The term "vegan leather" itself has created a buzz in the fashion industry. However, it is still unclear if vegan leather can replace traditional leather in the long

S. S. Muthu, M. Ramchandani, *Vegan Alternatives for Leather*,
Sustainable Textiles: Production, Processing, Manufacturing & Chemistry,
https://doi.org/10.1007/978-3-031-65365-0_1

term. Therefore, there is a need to have a deeper understanding and assessment of these products.

The aim of writing this book is to bring out the following important information for the scientific fraternity:

- Outline the environmental impact of traditional leather.
- Discuss different types of vegan leather and their production process.
- Compare the environmental impact of traditional leather and the vegan leather.
- Address sustainability issues and environmental assessments within the vegan leather industry.
- Highlight production-related issues and challenges.
- Explore consumer perspectives and marketability within the luxury and fashion industry.
- Examine different product usages across various sectors and adoption.

2 Methodology

The research approach for this book is based on primary and secondary resources. Previous literature on the topic has been extensively consulted for the scientific and theoretical exploration of vegan leather alternatives. To deeply understand and assess the industry, we collected primary data through the interview technique. As industry knowledge is not easily revealed by companies, it is difficult to collect reliable and accurate data that is also recent. For this purpose, we contacted three different individuals for interviews to assess the challenges of the production process, the market and consumer demand, material sustainability, and the future of the vegan leather industry. Some details are as follows:

- **Mr. Adrian Lopez Velarde** is the co-founder of the vegan leather company Desserto from Mexico. The company produces vegan leather made from the cactus plant and is well known for its usability across the bags, footwear, furniture interiors, and automotive industries.
- **Ms. Hyunjoo Kim** is an artist based in South Korea. She specializes in handmade luxury products made with natural materials like Hanji paper. Her work is internationally recognized by art curators, museums, and the luxury interior industry.
- **Mrs. Min Zhou Carpentier** is the owner of ÉMZI Paris, a footwear brand from France. She uses traditional leather in her footwear collections and some vegan leather alternatives like rubber and linen for certain footwear collections.

The excerpts from the interviews are provided for assessment throughout the chapters where relevant. Additionally, our knowledge and expertise from the industry and research also contributed to the holistic understanding of the vegan leather industry in this book. Incorporating many secondary data sources, the book delves into LCA studies, examining impact values derived from a diverse range of articles. Through meticulous analysis, it contrasts the environmental footprints of traditional leather and different types of vegan leather, shedding light on their nuanced differences. The book meticulously outlines the production processes of traditional leather alongside the various methods employed in the creation of contemporary vegan leather alternatives.

3 Need for Transparency

In the realm of purchasing vegan leather, consumer confusion looms large. When confronted with tags boasting "vegan leather," many buyers find themselves adrift, unsure of what this truly entails. Is vegan leather truly cruelty-free and environmentally sound? Despite the noble intentions of supporting animal welfare and ecological preservation, consumers often lack the vital information needed to make informed choices. This ignorance extends to the composition of vegan leather, which, unbeknownst to many, may be derived from plastics, laden with toxic chemicals and ultimately nonbiodegradable. The absence of comprehensive education on the intricacies of vegan leather production exacerbates this confusion, leaving consumers grappling with a hazy understanding of their purchasing decisions (Ashley, 2021).

Several companies and brands across different industries are now using vegan leather alternatives to become more sustainable. However, a major issue is the lack of transparency in communication about the raw materials used. For example, the use of plastics in some vegan leather alternatives (like PU and PVC) make it less sustainable for the environment, but it is marketed as eco-friendly (Dreesmann, 2023). Consequently, this leads to greenwashing by the brands. Therefore, it is important to understand the different environmental issues emerging from these vegan leather alternatives and how these issues can be addressed. Different environmental parameters are assessed in this book to discuss the sustainability of these materials and their various applications.

3.1 Types of Leather

For the purpose and scope of this book, it is important to distinguish between the leather types and their derivations. Detailed issues pertaining to these leather types (vegan and non-vegan) are elaborated in the next chapters. See Fig. 1.1 for an overview of leather types.

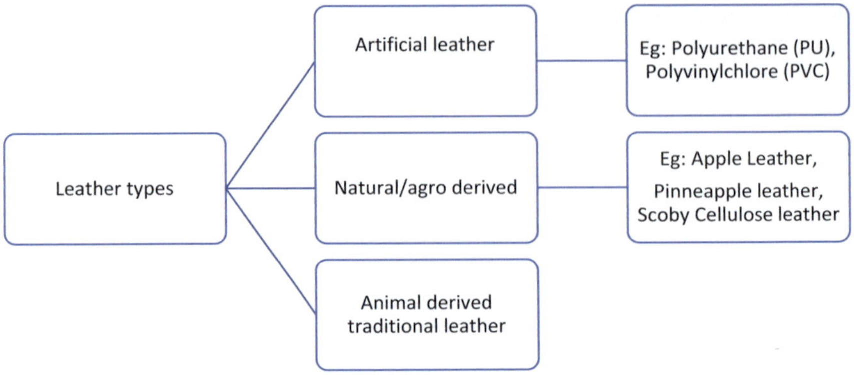

Fig. 1.1 Overview of leather types

3.1.1 Animal-Derived Traditional Leather

It is also referred to as genuine leather, which is derived from animal hides. Traditionally, it was only used by royals and the rich members of society. The animal skin was derived from slaughtering animals that were part of the food industry, like cows, buffaloes, pigs, goats, fish, and others. Products made from genuine leather include a variety of industry uses, such as bags, footwear, belts, furniture, clothes, automobile, and others. However, over the years, with the increase in demand for exotic leather from the luxury and fashion industry, several exotic animals like crocodiles, iguanas, snakes, and ostriches have been used.

3.1.2 Natural-Derived Alternatives to Leather

These constitute bio-based materials that substitute leather for different product uses and are derived from natural resources or agricultural industry. A combination of different raw materials and processes is involved in producing these types of vegan alternatives to leather. For example, MIRUM is a type of vegan leather made from cork, coconut husk, rice hull, natural rubber, vegetable oils, and minerals (NFW, 2021). Apple leather, also known as AppleSkin (Apple Skin, n.d.), is made from discarded apple skins and pomace residue from the apple industry. Piñatex is a type of vegan leather made from discarded pineapple leaves and fibers (Ananas Anam, n.d.). Symbiotic culture of bacteria and yeast (SCOBY) is a biofilm of cellulose, a by-product of the kombucha tea fermentation (Aadityawadi et al., 2021).

3.1.3 Artificial Leather

These are artificial or synthetically made leather alternatives (also known as faux leather), which are used for making upholstery, clothing, footwear, and other products (Wikipedia, 2023). For example, PVC or polyvinyl chloride is a type of artificial leather using a plasticizer called a phthalate for flexibility and softness. PU or polyurethane is made from a flexible polymer which is coated usually on a backing fabric like polyester (Dreesmann, 2023).

4 Government Norms and Global Standards

In Germany, the German Leather Federation (VDL) argues that the term "vegan leather" is misleading and anticompetitive because the traditional perception and reference of the term "leather" is animal-derived. Therefore, companies can be prosecuted for labeling their artificial leather as "vegan leather" or "eco leather," especially when derived mainly from plastic (VDL, 2023). Similarly, Portugal has passed legislation regulating the use of the term "leather" through the Leather Decree, Decree Law No. 3/2022, which prohibits the use of the word "cuoro" (leather in Portuguese) with prefixes, suffixes, and other adjectives that could mislead consumers (SGS, 2022).

According to the German designation regulation RAL 060 A2, the term "leather" is reserved for materials derived from animal hide and is distinguished from other materials. They also provide provisions for coated leather, composite materials, upholstery, etc. Additionally, the German Institute for Standardization, known as the Deutsches Institut für Normung (DIN), serves as an independent platform for standardization in Germany and worldwide, such as DIN EN 15987 for the leather trade (VDL, 2023). DIN is also involved in standardizing the evaluation of synthetic leather used in upholstery and its resistance to soiling through standards like DIN 16460–2 and DIN EN 15973:2011 (DIN, 2021; ANSI, n.d.).

The International Organization for Standardization (ISO), based in Geneva, Switzerland (ISO, n.d.), and European Standards (EN) are also used for standardizing leather used in upholstery, shoes, and various performance-related tests, e.g., ISO 16131 used in upholstery. Additionally, the European Textile Labelling Regulation (Regulation (EU) No. 1007/2011) also regulates the naming and labeling of non-textile parts of animal origin (VDL, 2023), such as fur and leather.

By the commonality of terms used within the industry and for the purpose of this book, we are referring to "vegan leather" as mainly those products that "look and feel" like the texture of genuine leather but are not derived from animals and those materials that can be used as alternatives for products currently using traditional leather.

The leather industry must adhere to government regulations and global standards aimed at minimizing environmental impact and fostering sustainable practices. Conformance to these regulations and standards is crucial for ensuring the industry's long-term sustainability and meeting the growing consumer preference for eco-friendly products. Regulatory bodies have enforced a spectrum of environmental compliance mandates that leather manufacturers are obliged to adhere to. These mandates commonly encompass constraints on the release of pollutants into the atmosphere and waterways, limitations on the utilization of dangerous substances, and guidelines for the appropriate handling and disposal of waste materials.

4.1 ISO

The International Union of Leather Technologists and Chemists Societies (IULTCS) serves as the global standard-setting organization tasked with creating international standards outlining testing methods for leather, excluding finished articles. Within the leather industry, ISO/TC 120 is the pertinent committee entrusted with standardization efforts. This committee focuses on standardization across various aspects of leather production, covering raw hides and skins, tanned materials, finished leather, and leather products (ISO 120, n.d.)

4.2 WTO

The World Trade Organization (WTO) has established an Agreement on Technical Barriers to Trade, which encompasses various products, including leather, and is designed to uphold specific standards during their international export, import, or trade activities among nations (WTO, 2024).

4.3 Certifications

Certifications like BeVeg, V-Label, Certified Vegan, Leather Working Group (LWG), Sustainable Leather Foundation (SLF), The Vegan Society, Italian Certification Institute for Leather (ICEC), OEKO-TEX, and similar organizations are pivotal in the industry. They evaluate different facets of leather production, such as chemical handling, environmental protocols, waste disposal, social assessments, and traceability measures (BeVeg, n.d.; V-Label, n.d.; Leather Working Group, n.d.; Sustainable Leather Foundation, n.d.; The Vegan Society, n.d.; Italian Certification Institute for Leather, n.d.; OEKO-TEX, n.d.).

4.4 Global Benchmark Standards

Textile Exchange has established minimum environmental and social criteria that standards and certification schemes must uphold. Suppliers must adhere to a certified standard meeting these criteria to participate in the Leather Impact Accelerator

(LIA). All standards and certifications endorsed by the LIA undergo third-party assessment to ensure they meet the established benchmark (Textile Exchange, n.d.).

5 Challenges

5.1 Scientific Challenges

Traditional leather and its vegan alternatives are mostly tested and compared based on several attributes. For example, the study by Meyer et al. (2021) elaborates on the performance of vegan leather alternatives based on their mechanical properties such as tensile strength, flexibility, biodegradability, tear resistance, water vapor permeability, and absorption. Their study further investigated the material composition, structure, and tactile properties of vegan leather. The major challenge that remains unanswered is the physical advantage of vegan leather alternatives compared to traditional leather, for which a LCA of these materials would be necessary (Meyer et al. 2021).

5.2 Marketing and Product Commercialization

As there is an increase in market demand and brands experimenting with vegan leather products, the market is expected to be worth $90 billion by 2025 (Hirsh, 2020). Many products, such as shoes, bags, upholstery, watches, etc., are being adopted by companies to promote the vegan leather alternatives. However, it is unclear how well the supply and availability of these materials are managed and at what prices. Some questions we aim to assess include the following: Is it cheap or expensive to use vegan leather in the fashion industry? Are retailers interested in selling vegan leather? And how can sustainable vegan alternative products be effectively marketed?

Manufacturers of traditional leather question the long-term usability and multi-scale applicability of vegan leather. They believe that the superiority of traditional leather has stood the test of time and will continue to be superior to vegan leather alternatives. On the other hand, consumers are increasingly receptive to innovation in the fashion industry. However, innovative vegan alternatives often lack communication to properly inform consumers about their production processes and distinctive attributes that satisfy consumer demand for sustainable products. Although some brands within the luxury industry are embracing the use of vegan leather, many are still reluctant to completely replace traditional leather with it.

The subsequent chapters in this book will detail these problems and challenges.

References

Aditiawati, P., Dungani, R., Muharam, S., Sulaeman, A., Hartati, S., Dewi, M., & Rosamah, E. (2021). *The nanocellulose fibers from symbiotic culture of bacteria and yeast (SCOBY) Kombucha: Preparation and characterization*. IntechOpen. https://doi.org/10.5772/intechopen.96310.

Ananas Anam. (n.d.). Retrieved October 25, 2023, from https://www.ananas-anam.com/

ANSI. (n.d.). Retrieved October 25, 2023, from https://webstore.ansi.org/standards/din/dinen159732011

Apple Skin. (n.d.). Retrieved October 25, 2023, from https://www.appleskin.com/

Ashley, V. (2021). Knowledge of leather alternatives: An exploratory study: Implications for education. *Journal of Textile Science & Fashion Technology, 7*(4). https://doi.org/10.33552/JTSFT.2021.07.000668

BeVeg. (n.d.). Accessed from www.beveg.com, Accessed on 13 Apr 2023.

Certified Vegan. (n.d.). Retrieved April 13, 2024, from https://vegan.org

DIN. (2021). Retrieved October 25, 2023, from https://www.din.de/en/wdc-proj:din21:346911899

Dreesmann, M. (2023). *Was ist pu-leder—und warum sie es vermeiden sollten!* Retrieved October 25, 2023, from https://www.manuel-dreesmann.com/blogs/information/what-is-pu-leather-and-why-you-should-avoid

Hirsh, S. (2020). *Vegan Leather market projected to be worth nearly $90 billion in next 5 years*. Retrieved October 16, 2023, from https://www.greenmatters.com/p/vegan-leather-market-growing

ISO. (n.d.). https://www.iso.org/home.html

ISO 120. (n.d.). Retrieved April 13, 2024, from https://www.iso.org/committee/51974.html.

Italian Certification Institute for Leather. (n.d.). *The reference for leather sector certifications | ICEC*, Retrieved April 13, 2024, from https://www.icec.it/en,

Leather Working Group. (n.d.). *Leather manufacturer standard*, Retrieved April 13, 2024, from https://www.leatherworkinggroup.com/certification/leather-manufacturer-standard/

Meyer, M., Dietrich, S., Schulz, H., & Mondschein, A. (2021). Comparison of the technical performance of leather, artificial leather, and trendy alternatives. *Coatings, 11*(2), 226. https://doi.org/10.3390/coatings11020226

NFW. (2021). *How Mirum® is made*. Retrieved October 16, 2023, from https://blog.naturalfiberwelding.com/how-mirum-is-made.

OEKO-TEX. (n.d.). *OEKO-TEX® Leather Standard*, Retrieved April 13, 2024, from https://www.oeko-tex.com/en/our-standards/oeko-tex-leather-standard

Peta. (2022). *Veganes Leder: Das sind die 11 besten pflanzlichen Lederalternativen*. Retrieved October 16, 2023, from https://www.peta.de/veganleben/veganes-leder/#:~:text=Veganes%20Leder%20bezeichnet%20alle%20Lederimitate,Stoffe%20wie%20Ananas%2D%20oder%20Kaktusleder.

SGS. (2022). *Portugal's Leather decree comes into force*. Retrieved October 20, 2023, from https://www.sgs.com/en/news/2022/05/safeguards-05722-portugals-leather-decree-comes-into-force.

Stella McCartney. (2021). https://www.stellamccartney.com/gb/en/stellas-world/the-worlds-first-mylo-garments-created-from-vegan-mushroom-leather.html

Sustainable Leather Foundation. (n.d.). Retrieved April 13, 2024, from https://sustainableleatherfoundation.com/what-we-do

Textile Exchange. (n.d.). *Leather impact accelerator supports the supply chain to overcome its core challenges*. Retrieved April 20, 2024, from http://textileexchange.org/leather-impact-accelerator/.

The Vegan Society. (n.d.). Retrieved April 13, 2024, from https://www.vegansociety.com/

V-Label. (n.d.). Retrieved April 13, 2024, from https://www.v-label.com.

Wikipedia. (2023). *Artificial leather*. Retrieved October 16, 2023, from https://en.wikipedia.org/wiki/Artificial_leather

WTO. (2024). *Technical barriers to trade, 2024*. Retrieved April 13, 2024, from https://www.wto.org/english/tratop_e/tbt_e/tbt_e.htm

Chapter 2
Environmental Issues of Traditional Leather and Need for Vegan Leather

1 Introduction

Leather is one of the most versatile and highly useful materials in our daily life. It is too difficult to imagine daily life without leather materials. We use leather materials in many forms. As with any other industrial product, leather products also create environmental issues from cradle to grave. Just looking at recent statistics from Statista in 2023, the leather sector has a remarkable market size of over US$ 220 billion, generating a huge profit of over US$ 50 billion annually, with an expected exponential rise of $92.36 billion by 2028 (Statista, 2023). Additionally, the leather goods market is expected to reach $496.37 billion in 2027 at a CAGR of 7.06% (Leather Goods Global Market Report, 2023). Although it is used enormously across the globe, and there is strong demand continuing to grow for leather products, the leather industry also faces severe challenges on all fronts, including social, environmental, economic, and ethical.

As is well known, leather is valued for its durability and flexibility. Traditional animal leather is obtained from animal rawhides and skins (Duraisamy et al., 2016). An important process in producing leather products is tanning, which converts rawhides and skins into non-decaying material. This process is crucial to prevent leather from being attacked by bacteria, etc. and to make it withstand damage from chemical degradation and mechanical deformation. After tanning, leather gains desirable properties such as better hydrothermal stability, breathability, durability, and increased strength. One of the major attributes of leather products is their stability, achieved mainly through the strong interlocking of collagen fibers with the tanning agent applied during the tanning process. The tanning agent can be organic or inorganic, with inorganic agents mainly consisting of chromium salts that form

---

Subramanian Senthilkannan Muthu is the primary author of this chapter.

S. S. Muthu, M. Ramchandani, *Vegan Alternatives for Leather*,
Sustainable Textiles: Production, Processing, Manufacturing & Chemistry,
https://doi.org/10.1007/978-3-031-65365-0_2

coordination complexes with the skin collagen, thereby biochemically stabilizing it (Maina et al., 2019).

Tanning is an important process in leather processing, as mentioned earlier, and it can be classified into three substages:

- Pre-tanning.
- Tanning.
- Post-tanning.

There are various types of tannages available, and depending on the quality requirements of the leather, a suitable type of tannage can be chosen and applied. The various types of tanning processes are outlined below:

1. Mineral tanning: Also known as "chrome" tanning, this is the most extensive and common process, accounting for approximately 85% of tanned hides worldwide. This process involves soaking the hides in drums containing mineral tannins, primarily chromium sulfate.
2. Vegetable tanning: Another important processes in the leather sector, it holds around a 10% market share worldwide. This method uses natural vegetable tannins to modify the protein structure of the hide into leather.
3. Aldehyde tanning: This process produces leather known as "chrome-free" or "wet white" leather. Tanning agents such as formaldehyde, glutaraldehyde, and oxazolidine are used in this method.
4. Oil tanning: It is an older method where fish oil or other oils and fatty substances are used for tanning.
5. Synthetic tanning: Tanning is done using aromatic polymers such as novolac or Neradol, resulting in white-colored leather (Kesarwani et al., 2015).

Any product's environmental footprint comes from five of its life cycle phases: raw material extraction, manufacturing processes, transportation, consumer use, and disposal. The environmental footprint of supply chains and their potential for alleviation chiefly lies within the supply chain and its partners. Similarly, the consumption phase of any product, including leather goods, which encompasses consumer use and disposal (end-of-life), is entirely influenced by consumer behavior.

As noted earlier, the leather sector has seen significant growth, and it is worth noting that the global footwear market is a multibillion-dollar industry. As of 2022, the footwear market generated approximately $4800 million in revenue, according to statistics (Statistica). With the rapid growth of the population and the leather sector in relation to this exponential rise, significant impacts are expected from all life cycle phases of leather products, including emissions and pollution from the leather supply chain and waste at the end of the products' life. According to US EPA data from 2017, the leather sector alone produces millions of tons of waste annually (EPA). As is well known, the footwear industry raises numerous environmental concerns due to its extensive use of hazardous chemicals, emissions of various pollutants, and solid waste generated during production processes and at the end of the product life cycle. The environmental impact of the leather sector must be reduced by considering potential opportunities throughout all life cycle phases of leather products, from cradle to grave stages.

It is worth mentioning that material selection significantly influences the industry's environmental impact—specifically, the type of leather used in making leather products. There are possibilities to reduce environmental impacts throughout the supply chain, including changes in consumer behavior such as extended use and opting for environmentally beneficial alternatives like reuse and recycling instead of landfilling. To address the environmental issues associated with the entire leather industry, unconventional alternatives such as vegan leather are highly required.

This chapter is dedicated to addressing the environmental issues related to traditional leather production, highlighting the need for alternatives to traditional leather.

2 Environmental Issues of Traditional Leather Production

2.1 Production Processes and Their Environmental Concerns

Traditional leather production, from animal hide to finished product, is marked by significant environmental consequences. From the vast amounts of water used to the toxic chemicals in tanning processes, leather production raises critical questions about environmental and chemical sustainability. The process sequences of leather production and the corresponding environmental impact in each process are discussed in detail.

In general, the leather manufacturing process can be broken down into three phases:

A. Pre-tanning.
B. Tanning.
C. Post-tanning.

Pre-tanning is the preparatory stage where the collected rawhide is prepared for tanning. It includes a series of processes, namely (IL and FS Environment, 2010):

1. Preservation.
2. Soaking.
3. Liming.
4. Bating.
5. Degreasing.
6. Bleaching.
7. Pickling.
8. De-pickling.

Tanning is the next process in which the leather-making protein is made resistant to heat, enzymatic biodegradation, and thermomechanical stress. Tanning transforms rawhide into a stable material that dries into a flexible form without rotting and becomes suitable for a wide variety of final processes (Krishnamoorthy et al., 2012). The type of tanning is chosen based on the final application.

After tanning, leather undergoes dyeing for color, retanning for enhanced properties, setting to remove creases, drying to eliminate moisture, trimming for neatness, and finishing treatments for aesthetics and durability. These post-tanning processes are essential in refining the leather's appearance and characteristics before it is used in various products. The flowchart below (Fig. 2.1) clearly depicts the process flow in the traditional leather processing industry.

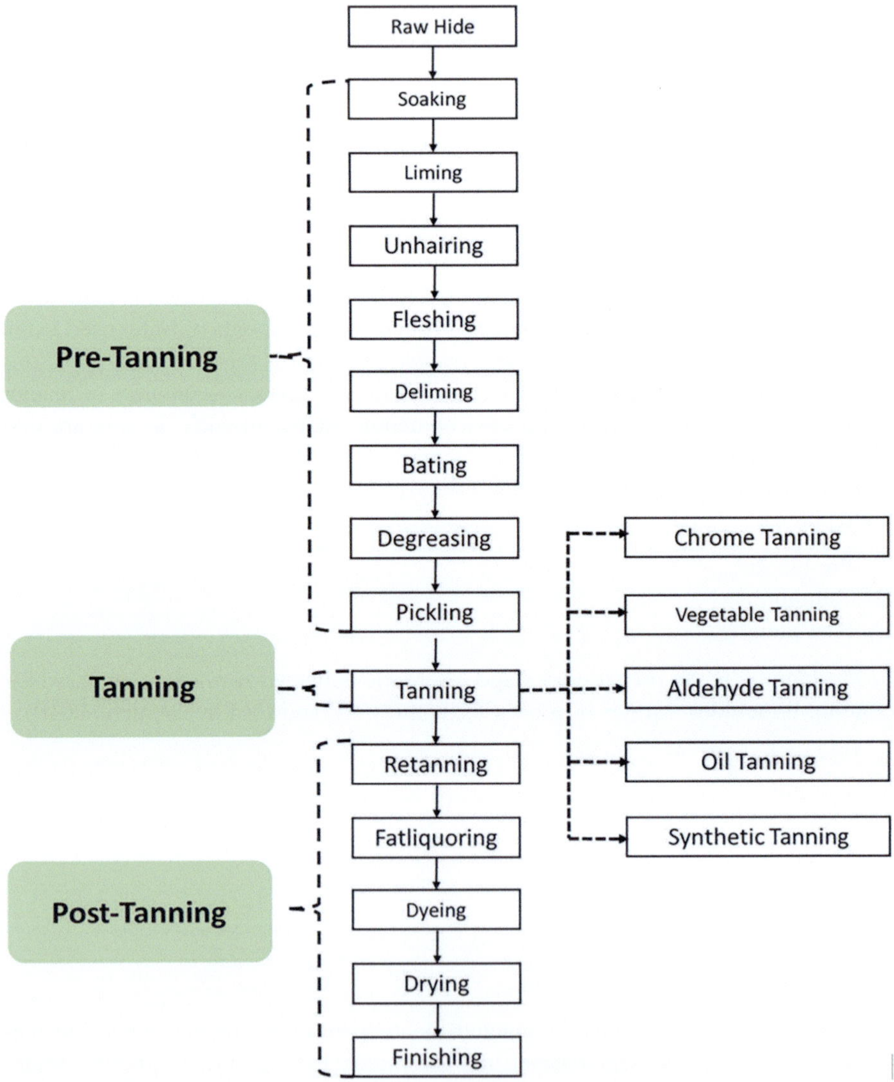

Fig. 2.1 Leather processing. (Source: Personal elaboration by the authors)

2.1.1 Pre-tanning

In leather production, hides and skins contain approximately 65% water and 30–35% proteins and fats. To prevent bacterial degradation caused by this high moisture content, it is necessary to reduce it to less than 30% (IL and FS Environment, 2010). Flayed skins and hides are preserved through drying or salting, with the goal of dehydrating them to resist putrefaction during transportation and storage. Preservation serves to destroy active bacteria, prevent bacterial activity, or avoid bacterial contamination. Sun-drying causes interfibrillary substances to coagulate, forming hard cementing substances that hinder fiber separation, making hides and skins challenging to rehydrate (Maina et al., 2019).

Soaking

Soaking, the initial step in leather processing, rehydrates skins to open fibers, removes curing salt and surface impurities, and facilitates subsequent treatments. Conducted in pits, paddles, or drums, it requires controlled conditions including pH (9–10), temperature (below 30 °C), and mechanical action to prevent fiber damage (Maina et al., 2019). Soaking duration varies from hours to days, influenced by the type of raw materials. Additives like surfactants, enzyme preparations, and bactericides may be included based on specific needs. The stages of soaking are dirt soaking, main soaking, and final Soaking (IL and FS Environment, 2010).

Liming and Unhairing

During the liming process in leather production, skins are treated with milk of lime (calcium hydroxide) and sharpening agents like sulfides to aid hair removal and keratinous matter. Liming raises the pH to 12–13, facilitating collagen fiber loosening, enhancing leather flexibility and fullness, and swelling the pelt to open the fiber structure. This process also aids in the splitting of the fiber structure, allowing better chemical penetration and more effective reactions. The degradation of hair results from the breakdown of disulfide links in the amino acid cystine, characteristic of keratin proteins.

Fleshing

Fleshing, a mechanical operation in leather processing, removes adhering flesh from the fresh side of the pelt post-flaying. Manual fleshing with a knife is typical for lighter pelts, while heavy pelts require a fleshing machine. Adequate fleshing ensures chemical penetration in subsequent processes.

Deliming

The purpose of the deliming process is to eliminate lime and other alkalies from the limed pelt, achieved through repeated washing or chemical treatment. Chemicals like ammonium chloride, ammonium sulfate, or weak acids such as boric acid and acetic acid are used for combined lime removal (Maina et al., 2019). Carbon dioxide can also be utilized as a deliming agent, neutralizing alkalinity. Deliming adjusts pH from 12–12.5 to 8–9, enhancing enzymatic activity. For soft leather, ammonium chloride is used, while ammonium sulfate is employed for hard leather (IL and FS Environment, 2010).

Bating

Bating is a crucial step in leather processing, yielding a finished product that is smooth, flat, flexible, soft, and stretchy. Proteolytic enzymes are added to open the fibrous structure of the pelt, making it softer, while also removing remaining lime. This process loosens scuds and eliminates unwanted proteins, increasing stretch. Bating de-swells swollen pelts and prepares them for tanning, conducted at optimal enzyme temperatures of 35–40 °C (Maina et al., 2019). It aims to render leather pliable and remove undesirable proteins, with protease enzymes like Palkobat and Palkocid commonly employed (Kesarwani et al., 2015).

Degreasing

Degreasing removes excess natural fats from the skin, preventing issues like fatty acid spues and uneven dyeing. Typically employing organic solvents, it is crucial for quality preservation, especially in sheepskin processing, where fats comprise 30–40% of raw weight.

Pickling

The pickling process adjusts collagen conditions for tanning agents by lowering pH with acids and salts, facilitating subsequent tanning reactions while preventing rapid skin-chromium compound combination. Common pickling agents include salts (5–10%) and acids (0.6–1.5%). In this process, water (80%), salts (8–10%), formic acid (0.28–0.3%), and sulfuric acid (0.75–2% based on thickness) are applied (IL and FS Environment, 2010). Salt prevents acid swelling, while acid protonates collagen carboxyl groups, optimizing reactivity for chrome tanning. After pickling, the skin can be stored for extended periods without deterioration (Kesarwani et al., 2015). For vegetable tanning, a pH of 4–4.5 is maintained, while for chrome tanning, a pH of 2.5–7.3 is targeted to prevent swelling, aiding dehydration for subsequent processing.

2.1.2 Tanning

The tanning process transforms rawhide into durable, non-putrefying leather by introducing crosslinks into collagen, ensuring stability. It relies on various methods like mineral and non-mineral tanning, chosen based on leather properties, cost, and raw material availability. Basification, using sodium bicarbonate, follows tanning to fix materials onto leather, enhancing stability and resistance. Shrinkage temperature, indicating tannage degree, varies with each method; chrome-tanned leather typically surpasses 100 °C, while vegetable-tanned hovers around 80–85 °C. Masking agents like sodium formate further modify metal complexes, reducing susceptibility to reactions, ensuring quality leather production (Maina et al., 2019). Table 2.1 clearly outlines the different types of tanning along with their pros and cons.

Table 2.1 Different types of tanning

Tanning method	Process	Result	Advantages	Disadvantages
Chrome tanning	Uses chromium sulfate and sodium sulfate in the tanning liquor	Produces wet blue semifinished leather	Yields more supple and pliable leather	Environmental hazards due to chromium compounds
Vegetable tanning	Relies on plant extracts for tanning	Leather free from heavy metal use	Environmentally friendly process	Less stable in water, prone to discoloration. Tends to shrink and become less supple when soaked and dried. Limited color possibilities
Aldehyde tanning	Uses glutaraldehyde or oxazolidine compounds for tanning	Results in wet white leather	Provides outstanding smoothness and water-absorbent potential	Hazardous due to the use of formaldehyde, posing health risks to workers
Synthetic tanning	Involves tanning with aromatic polymers such as novolac or Neradol	Produces white-colored leather	Developed as an alternative to vegetable tanning when tannins were scarce	Limited color range, may not provide the same aesthetic appeal as other tanning methods
Oil tanning	Employs oils to tan leather, often with animal fats or fish oils	Produces soft, flexible leather	Provides excellent water resistance and durability	Can be time-consuming and require special expertise. May impart odor to the leather

Source: Personal elaboration by the authors

2.1.3 Post-tanning

Post-tanning operations play a vital role in refining the quality and characteristics of leather. The process begins with sammying, a mechanized method aimed at removing excess moisture from wet blue leather. Following sammying, the material undergoes splitting, where it is divided into the desired thickness using a specialized machine. Subsequently, shaving and trimming are conducted to level the semifinished leather, eliminating irregularities and excess material to achieve uniformity. The process is followed by neutralization, retanning, dyeing, drying, and finishing.

Neutralization

Neutralization involves adjusting pH post-basification to eliminate acidity, ensuring thorough dye penetration and preventing leather deterioration. Common agents like sodium carbonate and sodium bicarbonate are used to achieve a pH range of 5.4–6.5. The process is crucial for preparing leather for subsequent treatments and enhancing its dyeing and finishing properties.

Retanning

Retanning utilizes retanning agents such as vegetable extracts, resins, and syntans to modify leather properties for desired characteristics like fullness, elasticity, and improved performance. These agents interact with the collagen matrix, either chemically or by filling voids between fibers, enhancing leather quality and facilitating downstream processes like dyeing and finishing.

Fatliquoring

Fatliquoring involves lubricating leather fibers with oils to prevent fiber sticking during drying, control softness, and improve strength. Various oils, including animal, vegetable, mineral, or synthetic oils, are applied in the form of emulsions to ensure uniform penetration and desired leather properties.

Dyeing

Dyeing encompasses coloring leather using different dye types like acid, direct, metal complex, and basic dyes. The dyeing process plays a crucial role in meeting fashion trends and environmental concerns.

Drying and Finishing

Drying and finishing involve removing excess water from leather to enhance texture and flexibility. Different drying methods like oven drying or vacuum drying are employed, followed by surface treatments such as coatings and finishes to cover defects and impart desired appearance and characteristics. Finishing materials like casein, nitrocellulose, and PU are mixed with natural substances to regulate color and texture.

2.2 Environmental and Chemical Issues in the Leather Supply Chain

The leather waste includes wastewater, air emissions, solid waste, and hazardous materials (IL and FS Environment, 2010). In research by Ozgunny et al., they have compared several data and concluded approximately 200 kg of leather is manufactured from 1 ton of wet salted hide. Leather factories dispose of over 600 kg of solid waste during the process of converting rawhide into leather. This waste primarily consists of protein and fat, making up more than 60% of the weight of the original rawhide. Unfortunately, these valuable materials are discarded into the environment without being utilized effectively (Ozgunny et al., 2007).

According to data presented in the 2020 Social and Environmental Report for the European Leather Industry, findings from surveys conducted in 2016 and 2018 revealed that European tanneries consumed an average of 2.15 kg of chemicals per square meter of finished leather (Social & Environmental Report, 2020).

Tanneries generate three main types of solid wastes. The first type is untanned solid tannery waste (USTW), which comprises hide trimmings and flesh. As these materials naturally decay, their environmental impact is minimal. The second type is tanned solid tannery waste (TSTW), which originates from operations conducted during or after the tanning process. This includes leather trimmings, shavings, dust, and small particles. The third type, tannery sludge (TS), is of particular concern. It consists of wastewater treatment chemicals and organic matter expelled from the hides during the tanning process (Famielec et al., 2020).

Figure 2.2 explains the amount of input and output flow in chrome tanning for 1 Mg of rawhide. Figure 2.3 clearly depicts the input and output of leather production from soaking to dry finishing.

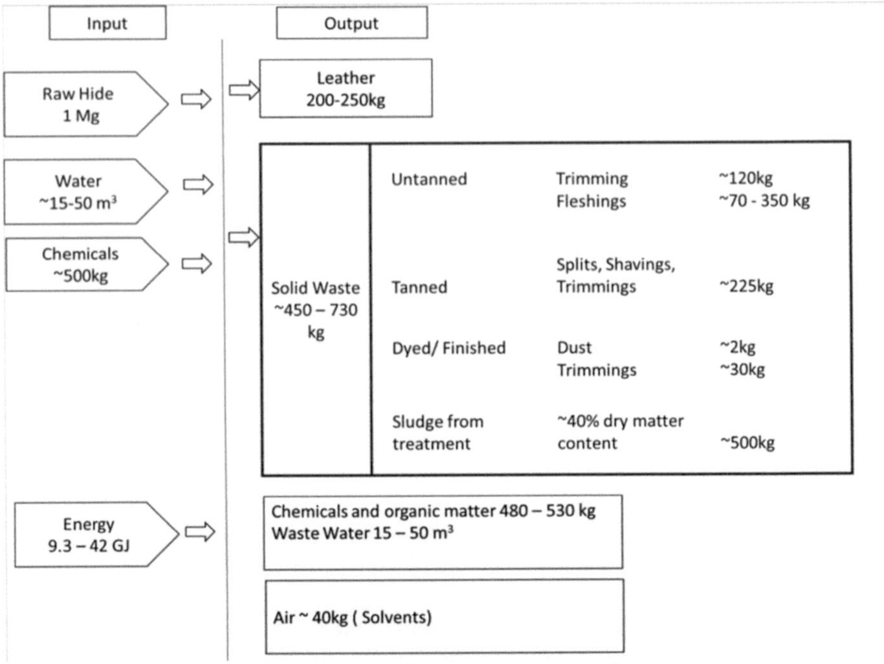

Fig. 2.2 Input and output overview of a conventional chrome tanning process for bovine salted hides per 1 Mg of rawhide processed. (Source: Famielec et al., 2020)

2.3 *Environmental Issues Related to Water and Wastewater from Leather Supply Chain*

2.3.1 Water Usage

Water is vital in tannery operations, with approximately 30–40 liters needed to process each kilogram of rawhide into finished leather (IL and FS Environment, 2010). Hides undergo soaking in large drums upon arrival, a process that consumes significant amounts of water. Some hides are treated with salt to prevent decomposition, further increasing water usage. In the beamhouse process, a 2019 study discovered water usage ranging from 7 to 25 cubic meter per ton of hides, while the tanning process typically utilizes between 1 and 3 cubic meter per ton of hides. Overall, the leather industry's reliance on freshwater puts pressure on local water supplies, estimated at 400 billion liters annually (Leather International).

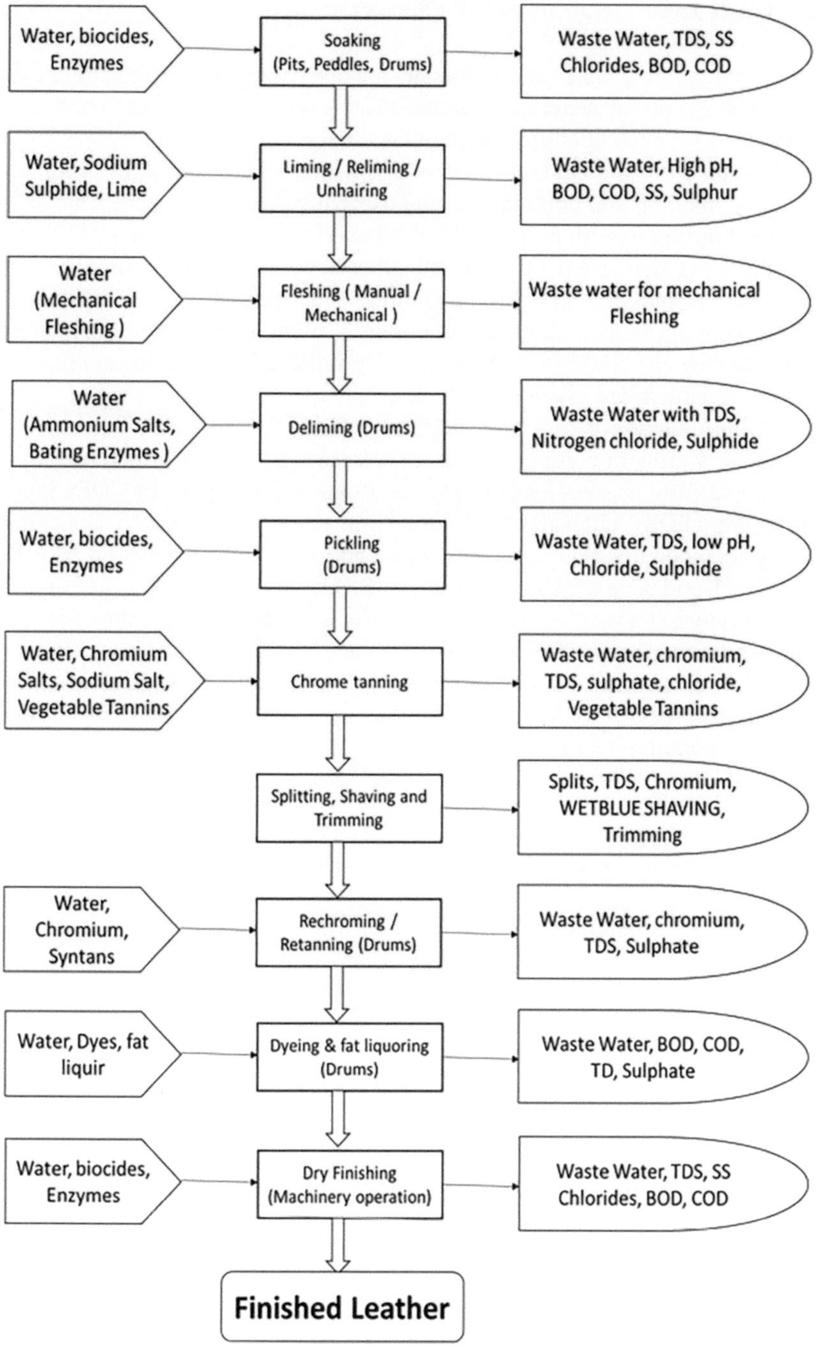

Fig. 2.3 Leather production flowchart along with input and output. (Sources: 23 and personal elaboration by the authors)

2.3.2 Wastewater Generation and Characteristics

Enormous amounts of water and pollutants are discharged during the entire tanning process. The details of water consumption for several steps and the characteristic pollutant load for each operation are tabulated below in Table 2.2.

Wastewater from the beamhouse process, which includes soaking, liming, and deliming, is alkaline and contains decomposing organic matter, hair, lime, sulfide, and organic nitrogen, with high levels of biochemical oxygen demand (BOD) and chemical oxygen demand (COD). On the other hand, wastewater from the tan yard process, such as pickling and chrome tanning, is acidic and colored. Effluent from vegetable tanning contains high organic matter, while chrome tanning wastes are characterized by elevated levels of trivalent chromium. The wastewater characteristics before treatment are given in Table 2.3.

The pollution load for 1 ton hide is given in Table 2.4.

The tannery effluent composition process-wise is given below in Table 2.5.

The pre-tanning process leads to pH variations and increases in COD, total dissolved solids (TDS), chlorides, and sulfates in tannery wastewater. Conventional dehairing methods involving sodium sulfide and lime contribute to 84% of BOD, 75% of COD, and 92% of suspended solids (SS) from tanneries. The use of sodium sulfide not only has adverse environmental effects but also impacts the efficiency of effluent treatment plants. The inefficient uptake of 50–70% chromium during commercial chrome tanning results in material wastage and ecological imbalances. Post-tanning processes also lead to significant modifications in TDS, COD, and heavy metal pollution (Sumita Dixit et al., 2015).

Table 2.2 Quantity of wastewater generated and pollution load of each step during processing 1 ton of skin/hide

	Process operations [Load kg/ton hide]					
Pollution load	Soaking	Unhairing/ Liming	Deliming and bating	Chrome tanning	Post-tanning	Finishing
Wastewater generated (kiloliter or m^3)	9–12	4–6	1.5–2	1–1.5	1–1.5	1–2
Suspended solids	11–17	53–97	8–12	5–10	6–11	0–2
COD	22–33	79–122	5–9	7–11	24–40	0–5
BOD	7–11	28–45	5–9	2– 4	8–15	0–2
Chromium	–	–	–	2–5	1–2	–
Sulfides	–	3.9–8.7	0.1–0.3	–	–	–
NH_3-N	0.1–0.2	0.4–0.5	2.6– 3.9	0.6–0.9	0.3–0.5	–
Total Kjeldahl nitrogen	1–2	6–8	3–5	0.6– 0.9	1–2	
Chlorides	85–113	5–15	2–4	40–60	5–10	–
Sulfates	1– 2	1–2	10–26	30– 55	10–25	–

Source: Sumita Dixit et al. (2015)

Table 2.3 Wastewater characteristics of leather processing before treatment

S. no	Pollution parameters	Average concentration in mg/liter (before treatment)	Average concentration in mg/liter (after treatment)
1	BOD	1850	700
2	COD	4500	3000
3	Chlorine (Cl)	5500	1200
4	SS (suspended solids)	3750	1500
5	Chromium (Cr)	165	38

Source: GCPC

Table 2.4 Pollution load to process 1 ton hide

S. no	Pollution parameter	Pollution load (kg)
1	Volume (cubic meter)	40
2	BOD	70
3	COD	180
4	Chlorine (cl)	270
5	Dissolved solids	600
6	SS (suspended solids)	100
7	Sulfides (S)	4
8	Chromium (Cr)	30

Source: GCPC

2.4 *Toxicity of Chemicals Used in Leather Industry*

A diverse range of chemicals is employed to transform leather into usable forms for various products. Table 2.6 outlines the purposes, chemical structures, LD50 values, and toxicity levels of these chemicals utilized in the leather industry. Regulatory bodies have imposed limitations on the utilization of these chemicals within the leather industry.

In leather manufacturing, hazardous chemicals are commonly used, including chromium, formaldehyde, azo dyes, sodium sulfide, chlorinated paraffin, heavy metals, perfluorinated compounds, and solvents. Chromium, extensively used in tanning, poses significant health risks due to the presence of hexavalent chromium, known for causing cancer and environmental pollution if not properly managed. Similarly, formaldehyde, present in leather processing as a preservative, is a known carcinogen, while azo dyes, employed for coloring leather products, can break down into toxic aromatic amines. Sodium sulfide, used in dehairing hides, can cause skin and respiratory issues, while chlorinated paraffin, utilized as a flame retardant, raises environmental concerns due to its persistence and potential health hazards. Additionally, heavy metals, perfluorinated compounds, and solvents used in leather manufacturing pose risks to both human health and the environment.

To mitigate these risks, strict safety measures, proper waste management, and adherence to regulations are essential. Initiatives focusing on alternative,

Table 2.5 Effluent composition

S. no	Parameter	Soaking	Beamhouse (liming, reliming, fleshing, deliming)	Pickling and chrome tanning	Wet finish (rechroming, dyeing, and fatliquor)	Composite (including washing)
1	Volume of effluent in liter/ton of hides/skin	6000–9000	6000–10,000	1500–3000	3000–5000	30,000–40,000
2	pH	7.5–8	8–12	2.2–4	3.5–4.5	7–9
3	BOD 5 days at 20 °C (total)	1100–2500	2000–8000	400–800	1000–2000	1200–3000
4	COD (total)	3000–6000	3000–15,000	1000–3000	2500–7000	2500–8000
5	Sulfide	–	50–200	–	–	30–150
6	TS (total solids)	35,000–55,000	6000–20,000	30,000–60,000	4000–10,000	15,000–25,000
7	DS (dissolved solids)	32,000–48,000	5000–15,000	29,000–58,000	3400–9000	13,000–20,000
8	SS (suspended solids)	3000–7000	3000–15,000	1000–2000	600–1000	2000–5000
9	Chloride	15,000–30,000	3000–6000	1500–25,000	500–1000	6000–9500
10	Total Cr	–	–	1500–3000	30–60	80–200

Source: GCPC
Note: Volume of wastewater applied for hides (cows and buffalo) and goatskins and not for sheepskins

eco-friendly tanning methods and responsible sourcing practices can help reduce reliance on hazardous chemicals and minimize environmental impact. Additionally, rigorous quality control measures and compliance testing are necessary to ensure consumer safety and regulatory compliance in the leather manufacturing industry.

2.5 *Solid Wastes and Leather Supply Chain*

Solid waste generated in the tannery industry includes salt from raw skin/hide dusting and raw skin/hide trimmings and hair containing lime and sulfides from the liming/dehairing process and fleshing from raw skins/hides. Additional solid waste includes wet blue shavings containing Cr2O3; wet blue trimmings from finishing processes containing chromium oxide (CrO), syntans, and dye; and buffing dust containing CrO, syntans, and dye. Tannery sludge's reducing characteristics

Table 2.6 Uses, structures, LD50, and toxicity of chemicals used within the leather industry

S. no	Name	Uses	LD50 in rats, oral (mg/kg)	Target organs
1	Benzyl butyl phthalate	Used in process to produce a microporous artificial leather coating/water vapor-permeable sheet materials	2330	Eyes, lungs, liver, reproductive system
2	Bis(2-ethylhexyl) phthalate (DEHP)	Used as plasticizers in processing shoe soles and artificial leather manufacturing	30,000	Liver and testes
3	Dibutyl phthalate (DBP)	Used as a phthalate plasticizer in the artificial leather industry	7499	Eyes, lungs, gastrointestinal tract, testes
4	Anthracene	Used as a tanning agent (for leather)	16,000	Kidney, liver, fat
5	Short-chain chlorinated paraffins (PBT)	Additive for leather treatment (renders smoothness to leather), leather clothing, and belts and as a leather oiling agent	3090	Liver, kidney, thyroid
6	Sodium dichromate	Principal raw material used in the production of chrome tanning materials for the leather industry like chrome tanning salts	Not available	Blood, kidneys, heart, lungs, eyes
7	Cobalt dichloride	Used in leather dyeing and finishing as well, found in tanned leather	80	Lungs, liver, kidney, heart, skin
8	Nonylphenol	Used in finishing	1475	Blood, lungs, eyes, skin, central nervous system, kidneys
9	Methylisothiazolinone	Biocide, microbiological protection	1800	Skin, eyes
10	N-methyl pyrrolidone	Coalescence, plasticizers, wetting agent	3914	Eyes, kidney, lymphatic system, liver, lung, testes
11	Formaldehyde	Leather finishing	100	Eyes, lungs
12	Heavy metals arsenic	Leather finishing	763	Liver, kidneys, skin, lungs, lymphatic system
13	Chromium	Used for dyeing	3250	Kidney, central nervous system, hematopoietic system

(continued)

Table 2.6 (continued)

S. no	Name	Uses	LD50 in rats, oral (mg/kg)	Target organs
14	Organotin compounds (Dibutyltin)	As a catalyst	175	Gastrointestinal tract, liver
15	Azo dyes	Used for dyeing	3418	Blood, liver, testes

Source: Dixit et al. (2015)

Table 2.7 Elemental composition of dry waste

Elements	Composition in dry waste (%)
C	43.6–48.2
O	20.8–23.0
N	13.8–15.1
H	5.8–6.2
S	1.0–2.0
Cr	2.6–3.9
Na	1.2–3.1
Si	0.5–1.9
Cl	0.4–2.1
Fe, Ca	0.5–1.0
Mg, K, Al	<0.5

Source: Famielec et al. (2020)

stabilize Cr (III) compared to Cr (VI) due to organic matter and sulfides. The considerable sludge produced by tannery plants impedes solid waste management due to the nonbiodegradability of tanned leather. Leather itself biodegrades slowly, and the treatment of various chemicals during the tanning process renders it resistant to chemical, thermal, and microbiological degradation (Sumita Dixit et al., 2015). The solid dry waste composition is given in Table 2.7.

2.6 *Emissions to Air from Leather Processing*

The emissions from leather processing include ammonia, hydrogen sulfides, and many other volatile organic compounds (VOCs). The emissions to air and their sources are detailed in Table 2.8 below.

Hydrogen sulfide emissions can cause acid rain and ocean acidification. The gas can lead to numerous health problems, including headache, pulmonary edema, unconsciousness, and even death (Sikander et al., 2021). VOCs mentioned in Table 2.7 include formaldehyde, which is a precursor in the formation of smog and poses potential health risks such as eye, nose, and throat irritation, headache,

Table 2.8 Emission to air from leather processing

S. no	Emission to air	Sources
1	Organic solvents	Degreasing, finishing
2	VOCs	Spray-finish machines, dryers
3	Sulfides	Beamhouse and effluent treatment
4	Ammonia	Beamhouse, deliming, dehairing, drying after dye penetration
5	Hydrogen disulfide	Beamhouse operations, effluent treatment plant (ETP) collections, ETP primary treatment unit
6	Methane	ETP anaerobic lagoons

Source: IL and FS Environment (2010)

Table 2.9 Total energy consumption, production, and specific energy consumption for the period 2018–2020

Year	Energy source	Energy consumption (MJ)	Production (sq. m)	Specific energy consumption (MJ/sq. m)
2018	Natural gas	5.31 × 107	8.03 × 105	90.65
	Electricity	1.97 × 107		
	Total	7.28 × 107		
2019	Natural gas	5.36 × 107	7.15 × 105	102.53
	Electricity	1.97 × 107		
	Total	7.33 × 107		
2020	Natural gas	3.21 × 107	3.7 × 105	116.15
	Electricity	1.09 × 107		
	Total	4.30 × 107		

Source: Morera et al. (2023)

nausea, and damage to the liver, kidney, and central nervous system damage and is found to be carcinogenic (Marsal et al., 2018).

2.7 *Energy and Greenhouse Gas Emissions from Leather Processing*

Traditional leather processing is an energy-intensive process that requires a significant amount of energy for heating and cooling water. This energy usage directly contributes to greenhouse gas emissions. The total energy consumption in a tannery is detailed in Table 2.9. The average specific energy consumption is 103 MJ/m^2 .

3 Available Tools and Techniques for Assessment

LCA (life cycle assessment) is a tool to quantify the environmental impact of a product, process, or service by assessing its environmental implications throughout its lifecycle. There are four stages to conduct an LCA as mentioned below (ISO 14044):

1. Goal and scope definition.
2. Inventory analysis.
3. Impact assessment.
4. Interpretation of results.

An LCA study conducted by Andrea Luca Tasca and Monica Puccini, which serves as an example here, explores a comprehensive system encompassing all stages from wet blue retanning to leather dyeing (crust leather) (Tasca & Puccini, 2019). The life cycle inventory (LCI) of the study is detailed in Table 2.10, and the impact details are provided in Table 2.11.

For a better understanding of the LCA process, a case study is presented that exemplifies the methodology of conducting LCI, encompassing every process involved in leather processing from rawhide to finished leather.

Case Study
UD PKX is a cow leather tanning industry that uses chromium as its tanning agents, located in Magetan District, Magetan Regency, East Java, Indonesia. UD PKX processes tons of raw cowhides into 285.4 kg of tanned cow leather. The LCA conducted by them is explained in detail below to understand the leather LCA (Ulya et al., 2021).

3.1 Life Cycle Assessment of Tanned Cow Leather

3.1.1 Goal and Scope Definition, Functional Unit, and System Boundaries

The LCA focuses on the chrome tanning process and is analyzed using a "gate-to-gate" perspective. The study covers the process from rawhide to finished leather, specifically chrome tanning. The functional unit is 285.4 kg of tanned cow leather. This quantity was chosen because it represents one batch production input, which is 1500 kg (Ulya et al., 2021). The system boundary starts from raw cowhide and ends with tanned cow leather, as detailed in Fig. 2.4.

Inventory Analysis (LCI)
All input and output data are collected from the production site. Input data includes water, electricity, and chemicals used, while output data includes collected wastewater, solid waste, and emissions (Ulya et al., 2021).

Table 2.10 Retanning, fatliquoring, and dyeing: inputs and outputs, referred to 1 kg of crust leather

Process	Input and output	Amount	Unit
Retanning	**Input**		
	Ethoxylated surfactant	0.003	kg
	Acetic acid	0.003	kg
	Sodium formate	0.005	kg
	Basic chromium sulfate	0.04	kg
	Water (35 °C)	3	kg
	Electric energy	1.75	kWh
	Heat	0.051	kWh
	Output		kg
	Wastewater	1.5	kg
Fatliquoring	**Input**		
	Sodium bicarbonate	0.012	kg
	Sodium formate	0.008	kg
	Syntan (tannin)	0.05	kg
	Sulpho chlorinated paraffin/epoxidized vegetable oil	0.1	kg
	Water (30 °C)	0.7	kg
	Water (50 °C)	1	kg
	Electric energy	0.55	kWh
	Heat	0.042	kWh
	Output		
	Wastewater	2.7	kg
Dyeing	**Input**		
	Formic acid	0.033	kg
	Azo dye	0.03	kg
	Cold water	0.3	kg
	Water (55 °C)	2	kg
	Electric energy	0.35	kWh
	Heat	0.08	kWh
	Output		
	Wastewater	2.8	kg

Source: Tasca and Puccini (2019)

Impact Assessment (Life Cycle Impact Assessment or LCIA)

LCIA is a qualitative and quantitative method used to explain the environmental impacts identified by the inventory. In the article, openLCA version 1.6.3 was used for performing the impact assessment. The database used is open_lca_methods_1_5_6.zolca.

Result Interpretation

The interpretation phase signifies the culmination of analyzing LCI data, encapsulating the refinement, finalization, and conclusion of various solution aspects alongside the impact derived from the analyzed data.

Table 2.11 Impact of retanning, fatliquoring, and dyeing

Impact category	Unit	Retanning	Fatliquoring	Dyeing
Climate change	kg CO2 eq	1.19	0.74	0.71
Ozone depletion	kg CFC-11 eq-10	9.72	7.67	8.33
Terrestrial acidification	kg SO, eq-10	4.9	3.15	2.72
Freshwater eutrophication	kg Peq-104	2.57	1.15	1.48
Marine eutrophication	kg N eq-10	2.21	1.68	3.01
Human toxicity	kg 1.4-DB eq-10-1	1.95	1.09	1.43
Photochemical oxidant formation	kg NMVOC-10-1	3.05	2.03	1.92
Particulate matter formation	kg PM_{10} eq-10	1.43	0.96	0.92
Terrestrial ecotoxicity	kg 1,4-DB eq-105	3.7	2.41	2.87
Freshwater ecotoxicity	kg 1,4-DB eq-10-3	4.54	2.56	4.34
Marine ecotoxicity	kg 1,4-DB eq-10	4.76	2.62	3.54
Ionizing radiation	kBq U235 eq-10-	1.91	0.92	0.76
Water depletion	m^3–10-2	3.95	1.57	1.71
Metal depletion	kg Fe eq-10-2	2.61	1.33	1.63
Fossil depletion	kg oil eq-10-1	3.32	2.74	1.92

Source: Tasca and Puccini (2019)

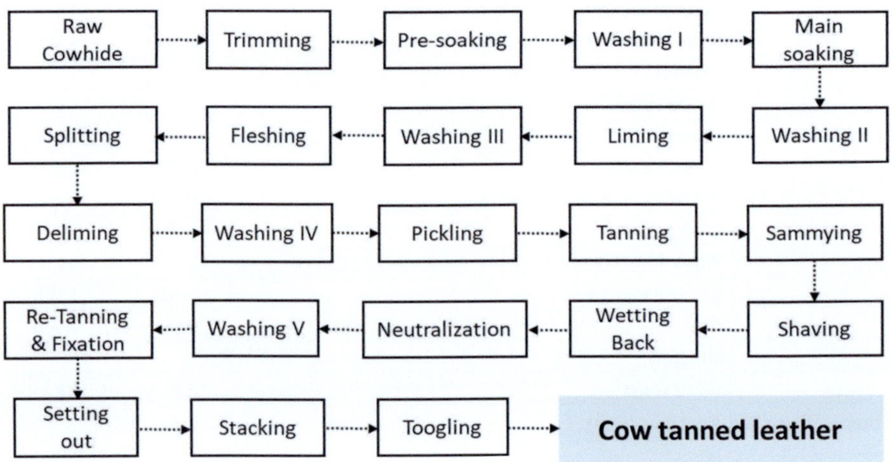

Fig. 2.4 System boundary. (Source: Ulya et al., 2021)

Results and Discussion

LCI

LCI data used to estimate the environmental impact at each stage of the leather processing is tabulated in Table 2.12.

Table 2.12 Inputs and outputs during tanning

S. no	Process	Inputs and outputs	Amount	Unit
1	Trimming	**Inputs**		
		Raw cowhide	1500	kg
		Outputs		
		Rejected cowhide	120	kg
		Raw cowhide	1380	kg
2	Presoaking	**Inputs**		
		Alcohols, C12–14, ethoxylated, unspecified	2.76	kg
		Alkylbenzenesulfonic acid, unspecified	2.76	kg
		Electricity	8.52	kWh
		Raw cowhide	1380	kg
		Water, ground	2415	kg
		Outputs		
		CO2	6.3	kg
		Cowhide	1380	kg
		Water, waste	2420.5	kg
3	Washing I	**Inputs**		
		Electricity	1.79	kWh
		Raw cowhide	1380	kg
		Water, ground	1380	kg
		Outputs		
		CO2	1.32	kg
		Cowhide, washing I	1380	kg
		Water, waste	1380	kg
4	Main soaking	**Inputs**		
		Electricity	43.51	kWh
		Raw cowhide, washing I	1380	kg
		Sodium carbonate	4.14	kg
		Water, ground	2760	kg
		Outputs		
		CO2	32.2	kg
		Cowhide, main soaking	1380	kg
		Water, waste	2764	kg
5	Washing II	**Inputs**		
		Electricity	2.2	kWh
		Cowhide, main soaking	1380	kg
		Water, ground	2415	kg
		Outputs		
		CO2	1.63	kg
		Cowhide, washing II	1380	kg
		Water, waste	2415	kg

(continued)

Table 2.12 (continued)

S. no	Process	Inputs and outputs	Amount	Unit
6	Liming	**Inputs**		
		Alcohols, C12–14, ethoxylate, unspecified	1.38	kg
		Electricity	14.27	kWh
		Glucose	2.76	kg
		Cowhide, washing II	1380	kg
		Limestone, in-ground	55.2	kg
		Sodium sulfide	1.38	kg
		Water, ground	4140	kg
		Outputs		
		CO2	10.56	kg
		Cowhide, liming	1380	kg
		Water, waste	4200.72	kg
7	Washing III	**Inputs**		
		Electricity	2.34	kWh
		Cowhide, liming	1380	kg
		Water, ground	2760	kg
		Outputs		
		CO2	1.73	kg
		Cowhide, washing III	1380	kg
		Water, waste	2760	kg
8	Fleshing	**Inputs**		
		Cowhide, washing III	1380	kg
		Outputs		
		Cowhide, fleshing	1352	kg
		Solid waste (cowhide)	27.6	kg
9	Splitting	**Inputs**		
		Electricity	9.4	kWh
		Cowhide, fleshing	1352	kg
		Outputs		
		CO2	6.96	
		Cowhide, splitting	676.2	
		Cowhide, by-product	676.2	
10	Deliming	**Inputs**		
		Alcohols, C12–14, ethoxylate, unspecified	1.35	kg
		Ammonium sulfate	13.52	kg
		Electricity	15.22	kWh
		Protease enzyme	2.03	kg
		Cowhide, splitting	676.2	kg
		Water, ground	676.2	kg
		Outputs		
		CO2	11.26	kg
		Cowhide, deliming	676.2	kg
		Wastewater	693.11	kg

(continued)

Table 2.12 (continued)

S. no	Process	Inputs and outputs	Amount	Unit
11	Washing IV	**Inputs:**		
		Electricity	3.03	kWh
		Cowhide, deliming	676.2	kg
		Water, ground	1352.4	kg
		Outputs		
		CO2	2.24	kg
		Cowhide, washing IV	676.2	kg
		Wastewater	1352.4	kg
12	Pickling	**Inputs**		
		Electricity	3.43	kWh
		Formic acid	3.38	kg
		Cowhide, washing IV	676.2	kg
		Sodium chloride	67.62	kg
		Sulfuric acid	1.35	kg
		Water, ground	676.2	kg
		Outputs		
		CO2	2.54	kg
		Cowhide, pickling	676.2	kg
		Wastewater	748.55	kg
13	Tanning	**Inputs**		
		Chromium VI	47.44	kg
		Electricity	70.87	kWh
		Formic acid	3.38	kg
		Cowhide, pickling	676.2	kg
		Oils, unspecified	1.35	kg
		Sodium carbonate	10.14	kg
		Sodium chloride	67.62	kg
		Sulfuric acid	1.35	kg
		Water, ground	676.2	kg
		Outputs		
		CO2	52.44	kg
		Cowhide, wet blue, tanning	709.33	kg
		Wastewater	774.24	kg
14	Sammying	**Inputs**		
		Electricity	1.79	kWh
		Cowhide, wet blue, tanning	709.33	kg
		Outputs		
		CO2	1.32	kg
		Cowhide, sammying	425.6	kg
		Wastewater	283.73	kg
15	Shaving	**Inputs**		
		Electricity	4.77	kWh

(continued)

Table 2.12 (continued)

S. no	Process	Inputs and outputs	Amount	Unit
		Cowhide, sammying	425.6	kg
		Outputs		
		CO2	3.35	kg
		Cowhide, solid waste	127.68	kg
		Cowhide, shaving	297.92	kg
16	Wetting back	**Inputs**		
		Alkylbenzenesulfonic acid	0.3	kg
		Electricity	3.86	kWh
		Formic acid		
		Cowhide, shaving	297.92	kg
		Water, ground	299.71	kg
		Outputs		
		CO2	2.86	kg
		Cowhide, wetting back	297.92	kg
		Wastewater	299.71	kg
17	Neutralization	**Inputs**		
		Electricity	2	kWh
		Cowhide, wetting back	297.92	kg
		Sodium formate	4.47	kg
		Water, ground	297.92	kg
		Outputs		
		CO2	2	kg
		Cowhide, neutralization	297.92	kg
		Wastewater	302.39	kg
18	Washing V	**Inputs**		
		Electricity	1.24	kWh
		Cowhide, wetting back	297.92	kg
		Water, ground	297.92	kg
		Outputs		
		Cowhide, washing V	297.92	kg
		CO2	1.6	kg
		Water, waste	297.92	kg
19	Retanning and fixation	**Inputs**		
		Acrylic acid	5.96	kg
		Catechol	1	kg
		Chestnut, pyrogallol tannins	5.96	kg
		Electricity	48.26	kWh
		Formic acid	4.47	kg
		Cowhide, washing V	297.92	kg
		Melamine	5.96	kg
		Oils, unspecified	17.88	kg
		Water, ground	287.92	kg

(continued)

Table 2.12 (continued)

S. no	Process	Inputs and outputs	Amount	Unit
		Outputs		
		CO2	35.98	kg
		Cowhide, retanning and fixation	297.92	kg
		Water, waste	351.96	kg
20	Setting out	**Inputs**		
		Cowhide, retanning, and fixation	297.92	kg
		Electricity	612.99	kWh
		LPG	45	kg
		Outputs		
		CO2	1.8	kg
		Dinitrogen monoxide	0.0011	g
		Cowhide, setting out	285.4	kg
		Methane	0.0001	kg
		Solid waste	15.88	kg
21	Stacking	**Inputs**		
		Electricity	11.19	kWh
		Cowhide, setting out	285.4	kg
		Outputs		
		CO2	8.28	kg
		Cowhide, setting out	285.4	kg
22	Toggling	**Inputs**		
		Electricity	22.38	kWh
		Cowhide, setting out	285.4	kg
		Outputs		
		CO2	16.56	kg
		Cowhide, setting out	285.4	kg

Source: Ulya et al. (2021)

Impact Assessment

Impact assessment carried out using openLCA software with the CML baseline version 4.42015 method. Environmental impacts encompass several categories, including climate change, human toxicity, marine aquatic ecotoxicity, freshwater aquatic ecotoxicity, terrestrial ecotoxicity, and photochemical oxidation. The summary of the LCA results is presented in Table 2.13 below (Ulya et al., 2021).

Table 2.13 Summary of environmental impacts

Impact category	Process	Amount	Unit
Climate change	Presoaking	6.30E+00	kg CO_2 eq
	Washing I	1.32E+00	kg CO_2 eq
	Main soaking	3.22E+01	kg CO_2 eq
	Washing II	1.63E+00	kg CO_2 eq
	Liming	1.06E+01	kg CO_2 eq
	Washing III	1.73E+00	kg CO_2 eq
	Splitting	6.96E+00	kg CO2 eq
	Deliming	1.13E+01	kg CO_2 eq
	Washing IV	2.24E+00	kg CO_2 eq
	Pickling	2.54E+00	kg CO_2 eq
	Tanning	5.00E+01	kg CO2 eq
	Sammying	1.32E+00	kg CO_2 eq
	Shaving	3.35E+00	kg CO_2 eq
	Wetting back	2.86E+00	kg CO_2 eq
	Neutralization	2.00E+00	kg CO2 eq
	Washing V	1.01E+00	kg CO_2 eq
	Retanning and fixation	3.60E+01	kg CO_2 eq
	Setting out	2.86E+02	kg CO_2 eq
	Stacking	8.28E+00	kg CO_2 eq
	Toggling	1.66E+01	kg CO_2 eq
Human toxicity	Tanning	9.26E+01	kg 1,4-dichlorobenzene eq
Marine aquatic ecotoxicity	Tanning	3.88E+04	kg 1,4-dichlorobenzene eq
Freshwater aquatic ecotoxicity	Tanning	3.12E+02	kg 1,4-dichlorobenzene eq
Terrestrial ecotoxicity	Tanning	1.03E-17	kg 1,4-dichlorobenzene eq
Photochemical oxidation	Setting out	600F-07	kg ethylene eq

Source: Ulya et al. (2021)

4 Conclusion

Traditional leather production poses significant environmental challenges, including deforestation for grazing land, greenhouse gas emissions from livestock farming and processing, water pollution from toxic chemicals used in tanning, excessive water consumption, and generation of waste. Vegan leather offers a more sustainable alternative by eliminating the need for animal agriculture, reducing resource usage, minimizing chemical inputs, and promoting innovation in eco-friendly materials. Thus, vegan leather helps mitigate the environmental impact of leather production while meeting consumer demand for ethical and cruelty-free products.

References

Cleaner Production Guidelines in Leather Industries, Gujarat Cleaner Production Centre. (2024). Accessed from https://www.pashudhanpraharee.com/wp-content/uploads/2022/12/CLEANER-PRODUCTION-GUIDELINES-IN-LEATHER-INDUSTRIES.pdf. Accessed on 13 Apr 2024.

Dixit, S., Yadav, A., Dwivedi, P. D., & Das, M. (2015). Toxic hazards of leather industry and technologies to combat threat: A review. *Journal of Cleaner Production, 87*. https://doi.org/10.1016/j.jclepro.2014.10.017

Duraisamy, R., Shamena, S., & Berekete, A. K. (2016). A review of bio-tanning materials for processing of fish skin into leather. *International Journal of Engineering Trends and Technology, 39*(1), 10–20.

EPA-Rubber and Leather: Material-Specific Data. (2024). Accessed from https://www.epa.gov/facts-and-figures-about-materials-waste-and-recycling/rubber-and-leather-material-specific-data. Accessed on 13 Apr 2024.

Famielec, S, et al. (2020). *Chromium concentrate recovery from solid tannery waste in a thermal process*. Project No. N R14 0028 06. EPA, https://doi.org/10.3390/ma13071533

IL and FS Environment. (2010). *Technical EIA guidance manual for leather/skin/hide processing industry*. Ministry of Environment and Forest, Government of India, Accessed from https://dste.py.gov.in/SEIAA/pdf/Sector/Leather_Industry.pdf. Accessed on 13 Apr 2024.

ISO 14044:2006. (2024). Accessed from https://www.iso.org/standard/38498.html. Accessed on 13 Apr 2024.

Kesarwani, P., Jahan, S., & Kesarwani, K. (2015). A review on leather processing. *International Journal of Applied Research, 1*, 977–982.

Krishnamoorthy, G., Sadulla, S., Sehgal, P. K., & Mandal, A. B. (2012). Green chemistry approaches to leather tanning process for making chrome-free leather by unnatural amino acids. *Journal of Hazardous Materials, 215*, 173–182. https://doi.org/10.1016/j.jhazmat.2012.02.046

Leather Goods Global Market Report. (2023). Accessed from https://www.globenewswire.com/news-release/2023/06/01/2680519/0/en/Leather-Goods-Global-Market-Report-2023.html. Accessed on 13 Apr 2023.

Leather International. (2024). *Water, waste and a wish for the future*. Accessed from https://www.leathermag.com/. Accessed on 13 Apr 2024.

Maina, P., Ollengo, M. A., & Nthiga, E. W. (2019). Trends in leather processing: A review. *International Journal of Scientific and Research Publications, 9*(12). https://doi.org/10.29322/IJSRP.9.12.2019.p9626

Marsal, A., Cuadros, S., Ollé, L., Bacardit, A., Manich, A. M., & Font, J. (2018). Formaldehyde scavengers for cleaner production: A case study focused on the leather industry. *Journal of Cleaner Production, 186*, 45–56.

Morera, J. M., Borri, E., Zsembinszki, G., Vérez, D., Gasa, G., Bartolí, E., & Cabeza, L. F (2023). Energy assessment of a tannery to improve its sustainability. *Sustainability, 15*(6):5599. https://doi.org/10.3390/su15065599

Ozgunny, H., Colak, S., Mutlu, M. M., & Akyuz, F. (2007). Characterization of leather industry wastes. *Polish Journal of Environmental Studies, 16*(6), 867–873.

Sikander, M., Kumar, L., Naqvi, S. A., Arshad, M., & Jabeen, S. (2021). Sustainable practices for reduction of environmental footprint in tanneries of Pakistan. *Case Studies in Chemical and Environmental Engineering, 4*, 100161. https://doi.org/10.1016/j.cscee.2021.100161

Social & Environmental Report. (2020). *The European Leather Industry*. Accessed from https://www.euroleather.com. Accessed on 13 Apr 2024.

Statista. (2023). *Leather & related products*. Accessed from https://www.statista.com. Accessed on 13 Mar 2024.

Statista. (2024) *Export value of leather and leather products in India in financial year 2022*. Accessed from https://www.statista.com. Accessed on 13 Mar 2024.

Tasca, A. L., & Puccini, M. (2019). Leather tanning: Life cycle assessment of retanning, fatliquoring and dyeing. *Journal of Cleaner Production, 226*, 720–729. https://doi.org/10.1016/j.jclepro.2019.03.335

Ulya, M., Arifuddin, A. L., & Hidayat, K. (2021). Life cycle assessment of cow tanned leather products. *IOP Conference Series: Earth and Environmental Science, 757*(1), 012066. https://doi.org/10.1088/1755-1315/757/1/012066

Chapter 3
Environmental Footprint, Toxicity, and Life Cycle Assessment Within the Vegan Leather Industry

1 Introduction

Conventional leather is a warm, breathable, durable, and flexible material made from animal skin. It is used to make clothes like jackets, shoes, belts, bags, and other accessories.

However, leather production comes with several environmental and ethical concerns, as discussed in the previous chapter. Harvesting leather from the skin requires farming. Farms house cattle for meat, dairy, and leather production. Some farms may cut corners with respect to the well-being of their livestock in order to produce large quantities of goods.

Additionally, the United Nations Food and Agriculture Organization reported that farming produces approximately 14% of all greenhouse emissions caused by human activity. This includes the traditional leather industry, which has a huge carbon footprint and is not environmentally sustainable (Reuters, 2024).

Leather making and tanning require a lot of resources, including energy and water, which can lead to waste. It uses chemicals like chromium salts, which create toxic wastewater that pollutes land and water alike. All these things are discussed in the previous chapter.

Vegan leather is easier to make and requires fewer resources than animal leather. Its production is energy- and water-efficient. Manufacturers make vegan leather in many ways and process it to resemble conventional leather in strength, texture, and appearance. Vegan leather requires low-cost natural fibers, agricultural waste products, and recyclable materials, making it both environmentally friendly and animal-friendly.

Vegan leather is a material made to look and feel like animal leather but without using any animal products. It is usually made from PU or PVC and sometimes from

---

Subramanian Senthilkannan Muthu is the primary author of this chapter.

S. S. Muthu, M. Ramchandani, *Vegan Alternatives for Leather*,
Sustainable Textiles: Production, Processing, Manufacturing & Chemistry,
https://doi.org/10.1007/978-3-031-65365-0_3

other plastics or even natural fibers like cotton. Many people choose vegan leather for ethical reasons, believing that it is wrong to use animals for our own purposes.

Here are a few of the benefits of vegan leather:

- Cheap, easy, and fast to make.
- Can be made using various sustainable materials.
- Highly versatile and available in many textures and finishes.
- Produced ethically and is environment- and animal-friendly.
- Stain-resistant and does not fade easily.
- Does not wear or crack easily.
- Water- and UV light-resistant.
- Has aided the growth of many sustainable fashion brands.

2 State of the Art

To know the advancements in vegan leather, the types are given below comprehensively:

Vegetable-tanned leather: Vegetable tanning is a traditional process that uses natural tannins derived from tree bark and other plant materials. This process eliminates the use of chromium, a toxic heavy metal used in conventional tanning, which has negative environmental and health impacts.

Vegetable-tanned leather is biodegradable and has a unique natural look and feel that many consumers find attractive.

Water-based leather finishing: Leather finishing is the process of treating the leather to achieve the desired look and feel. Traditionally, this process involves using solvent-based chemicals that are harmful to the environment and human health.

However, water-based leather finishing eliminates the use of these harmful chemicals, making it a more sustainable option.

Use of recycled leather: Leather waste is a significant problem in the industry, with millions of square feet of leather scrap generated each year. However, advancements in technology have made it possible to recycle leather waste and transform it into new leather products.

Recycled leather not only is eco-friendly but also has unique patterns and textures that add to its appeal.

Biodegradable leather: Researchers have been working on developing leather that is entirely biodegradable, which would significantly reduce the environmental impact of leather production. Some companies have already started producing biodegradable leather made from materials like mushrooms and other natural sources.

Chrome-free leather tanning: As mentioned earlier, chromium is a toxic heavy metal used in traditional leather tanning, which has negative environmental and health impacts. Chrome-free leather tanning eliminates the use of chromium and other harmful chemicals, making it a more sustainable option.

Digital printing on leather: Digital printing on leather allows for intricate designs and patterns to be printed without harmful chemicals. This process not only is eco-friendly but also allows for more customization and personalization in leather products.

Various research projects globally have focused on developing vegan leather alternatives, leading to the creation of innovative products. The vegan leather can be categorized as synthetic leathers and bioleather.

1. *Synthetic Leather.*
   (a) PU leather.
   (b) PVC leather.
   (c) Recycled synthetic leather.
2. *Bioleather.*
   (a) Fungal-based leather.
   (b) Bacterial-based leather.
   (c) Plant-based leather.
   (d) Animal-based leather.

3 Production of Vegan Leather

The production of vegan leather, utilizing plant-based, mushroom-derived, or synthetic materials, reduces harm to animals and environmental impact, reflecting a growing ethical consciousness driving sustainable alternatives globally.

3.1 Synthetic Leather

3.1.1 PU Leather

PU leather is created by immersing knitted or nonwoven fabric in a solution containing fossil plasticizer (i.e., polyurethane (PU)). This solution polymerizes around the fabric when submerged in dimethylformamide. Additionally, coating treatments are commonly applied to enhance the texture, durability, and appearance of the material (Pettersen & Krych, 2023). The flowchart of PU leather production is depicted in Fig. 3.1.

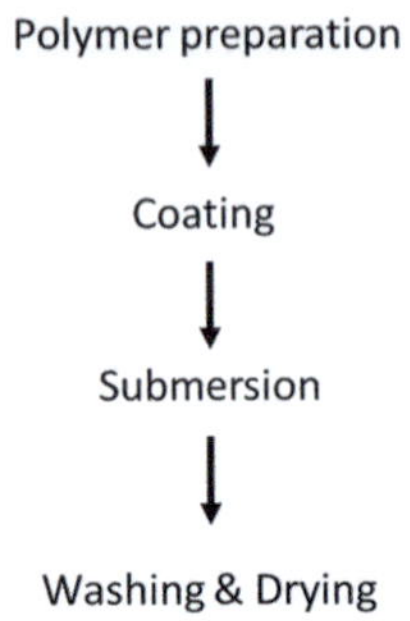

Fig. 3.1 PU leather manufacturing flowchart. (Source: Pettersen & Krych, 2023)

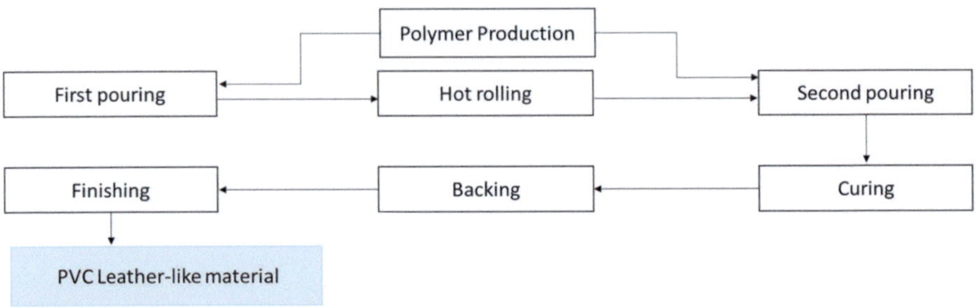

Fig. 3.2 Production of PVC leather-like material. (Source: Pettersen & Krych, 2023)

3.1.2 PVC Leather

In the production of synthetic leather, a solution comprising 55% fossil polymer (e.g., PVC), 40% plasticizer, and 5% filler is utilized to create a foamy layer that is bonded to a polyester fabric substrate. This process results in a durable material that mimics the texture of genuine leather. Following lamination, texturization methods are employed to enhance the surface, while coating treatments improve properties such as water resistance and durability. Despite offering a cruelty-free alternative, the use of fossil polymers prompts environmental considerations, driving ongoing efforts to develop eco-friendly alternatives in synthetic leather production (Gurera et al., 2018). The production overview of PVC leather is detailed in Fig. 3.2.

3.1.3 Recycled Synthetic Leather

Recycled synthetic leather is produced from postconsumer or postindustrial waste materials like recycled plastics or textiles (He et al., 2018).

3.2 *Bioleather*

3.2.1 Fungal-Based Leather

Mycelium-based vegan leather is produced by cultivating mycelium on sawdust and organic materials in a controlled indoor environment. The mycelium forms a foamy layer, akin to marshmallows, which is harvested, dehydrated, and pressed into thin sheets. These sheets undergo natural tanning and finishing processes to create the final mycelium-based vegan leather product. The production process minimizes waste by composting leftover by-products, resulting in an environmentally friendly and sustainable alternative to traditional leather (Ekegårdh & Kristersson, 2021).

The growth of mycelium is influenced by environmental factors such as temperature and humidity, which accelerate under favorable conditions. In the production process, mycelia spores are initially combined with nutrient-rich sawdust and spread across a large mat, where they develop into a dense, foamy mass. Any remaining residues are composted after mycelium collection. The resulting mycelium sheet is then treated and colored to produce Mylo™ material, a sustainable bioleather substitute utilized in various applications within the leather industry (Elkhateeb et al., 2022; Bolt Technology—Mylo, 2024).

Reishi™ represents a novel material brand crafted from high-quality mycelium, tailored for the creation of fashion and luxury goods. MycoWorks, a pioneering biotechnology firm, is closely associated with Reishi, driving innovation in sustainable materials for various industries (Reishi, 2024; Williams et al., 2023).

Mycelium-based leather was derived from the fruiting body of *Fomitella* spp. and *Phellinus ellipsoideus* (Elkhateeb et al., 2022). Figure 3.3 represents the production flowchart of fungi-derived leather. Figure 3.4 shows the mushroom mycelium fibers.

Meadow introduces Bio-VERA, a vegan engineered renewable alternative to leather. Microorganisms like yeast or bacteria are key in producing proteins and compounds for eco-friendly, highly efficient materials, offering a pathway to reduce ecological footprints across industries (Modern Meadow—Biomaterials, 2024).

Figure 3.5 outlines the performance, sustainability, and scalability comparison of conventional leather and Bio-VERA.

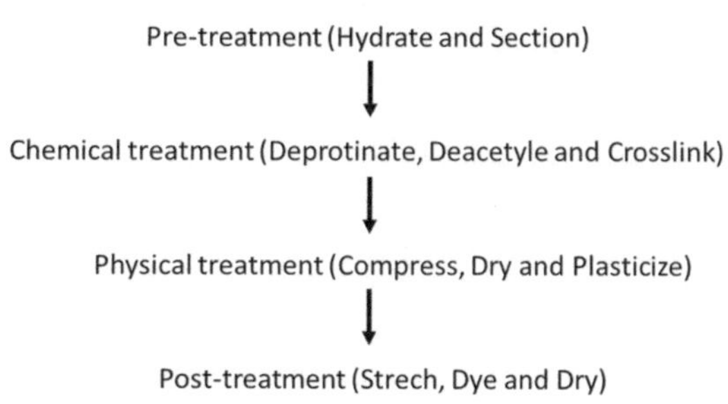

Fig. 3.3 Flowchart of fungi-derived leather. (Source: Pettersen & Krych, 2023)

Fig. 3.4 Mushroom mycelium fibers. (Source: Mushroom Leather—Watson & Wolfe, 2024)

3.2.2 Bacterial-Based Leather

Bacterial cellulose (BC) sourced from microorganisms like *Acetobacter xylinus* and Komagataeibacter presents a potential alternative to plant cellulose when produced at scale via biotechnology. BC, a pure cellulose biomaterial, resembles a gelatinous pellicle with excellent resilience, composed of three-dimensional interwoven nanofibers. Industrial synthesis of cellulose fibers from bacteria like Komagataeibacter, commonly found in kombucha tea fermentation, forms the basis for BC's potential as a leather substitute. These microbes create cellulose pellicles aerobically, either independently or in cooperation with other bacteria and yeasts, capable of producing flexible leather-like material with properties comparable to animal leathers. By utilizing industrial-scale fermentation of basic nutrients, significant amounts of BC pellicles could be generated, offering tactile characteristics akin to high-end, stretchy, finely finished leather. Studies have explored bioleather alternatives from kombucha-derived bacterial cellulose, resulting in composites with good shape stability and flexibility, featuring an average tensile strength and elastic modulus of 1.69 MPa and 100 MPa, respectively. The fabrication process involves culturing *Komagataeibacter xylinus* in various nutrient media, followed by harvesting and treatment of the resulting cellulose membranes. These membranes are then utilized to produce bio-based composite materials, which undergo characterization through scanning electron microscopy, Fourier-transform infrared spectroscopy, and thermal analysis to assess their structural and chemical properties (Kefale et al., 2023; Nguyen et al., 2021).

The appearance of the bacterial-based leather is shown in Fig. 3.6. Figure 3.7 depicts the process flow of Kombucha leather production.

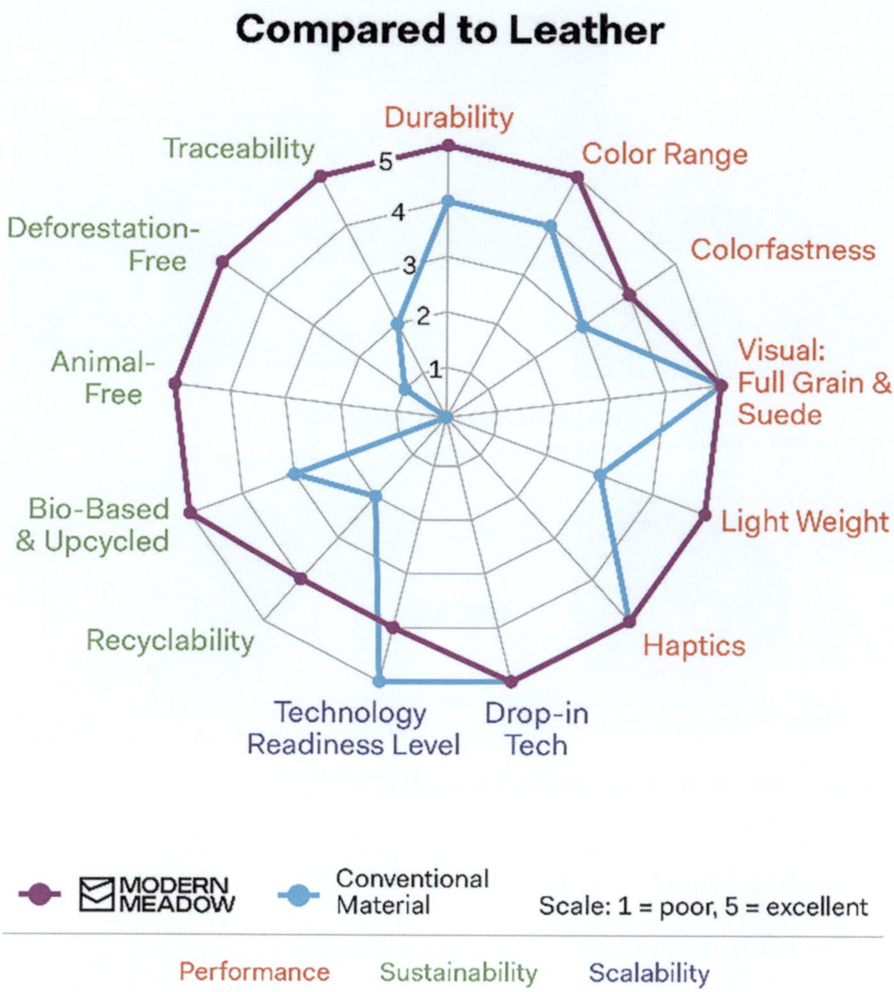

Fig. 3.5 Comparison of Bio-VERA and conventional leather. (Source: Modern Meadow—Biomaterials, 2024)

3.2.3 Plant-Based Leather

There are many types of plant-based leather which are listed below:

1. Pineapple leaf leather.
2. Cactus leather.
3. Apple leather.
4. Cork leather.
5. Grape leather.
6. Corn leather.
7. Banana leather.

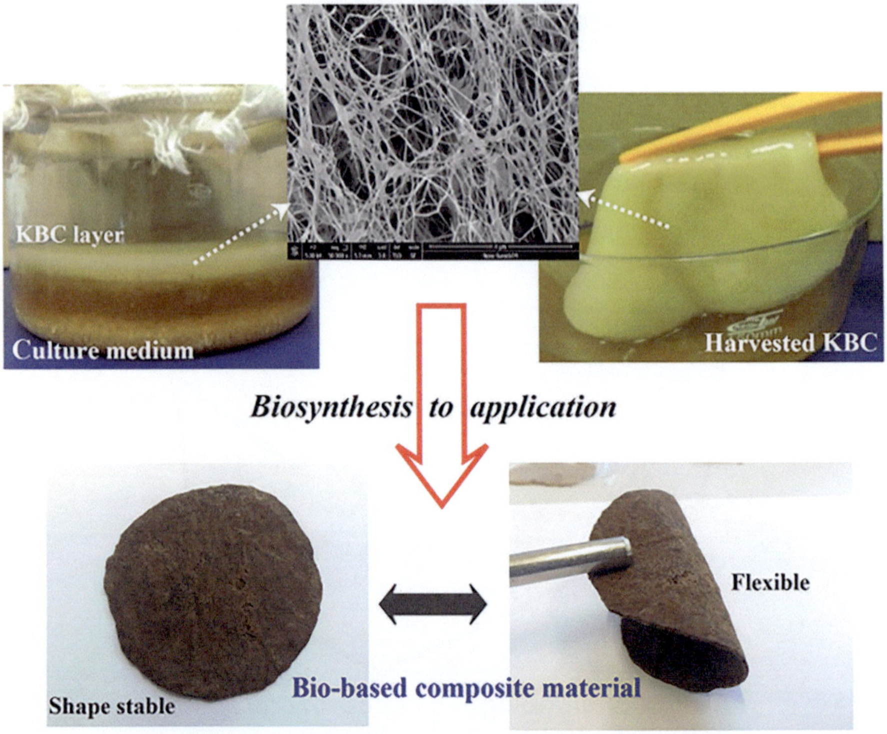

Fig. 3.6 Kombucha production. (Source: Nguyen et al., 2021)

8. Soyabean leather.
9. Bioleather1.
10. Coconut leather.
11. Orange silk leather.
12. Leaf leather.
13. Mango leather.
14. Hemp leather.

3.3 *Pineapple Leaf Leather*

The production techniques for various plant-based leathers vary depending on the source of cellulose utilized, with confidentiality often surrounding the businesses involved. Piñatex, a popular plant-based bioleather, is made from pineapple leaves' fibers, which are a by-product of the pineapple industry. The fibers are extracted,

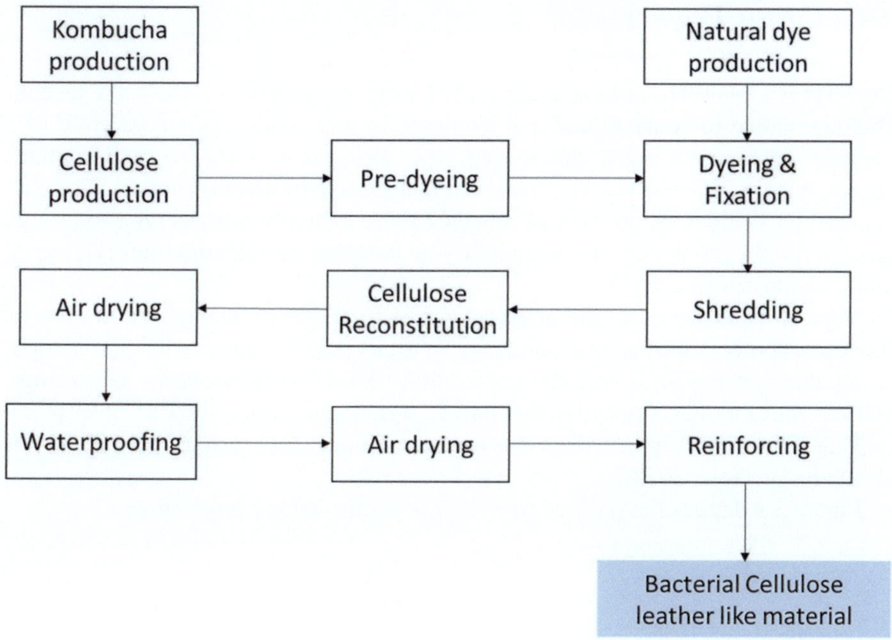

Fig. 3.7 Production flowchart of BC leather-like material. (Source: Pettersen & Krych, 2023)

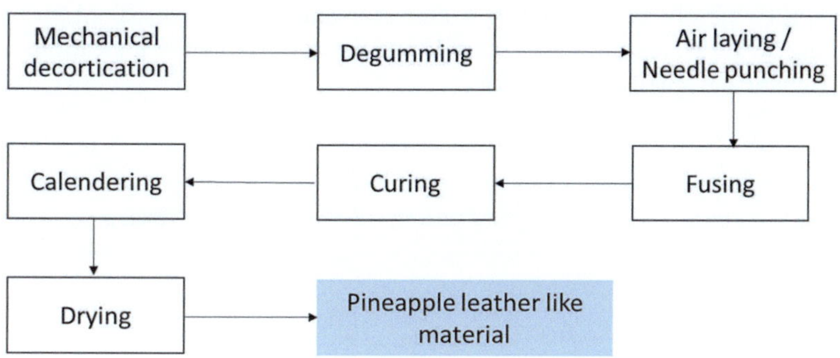

Fig. 3.8 Pineapple leather production. (Source: Pettersen & Krych, 2023)

cleaned, dried, and combined with polylactic acid (PLA) from maize to create a nonwoven mesh, which is then coated to produce the finished product known as Piñatex (Kefale et al., 2023; Gupta & Dave, 2021) (Anam Palf, 2024). The flowchart of pineapple leather production is depicted in Fig. 3.8.

3.4 Cactus Leather

Desserto is a plant-based bioleather crafted from cactus leaves, offering a sustainable alternative to conventional and synthetic leather. With layers consisting of a compact layer, foam layer, and woven base, Desserto is PVC-free and contains approximately 28% bio-polyurethane, reducing its environmental impact. The cactus used for Desserto is grown and collected twice annually, with its photosynthesis process further enhancing its sustainability by reducing overall emissions (Desserto, 2024; Gupta & Dave, 2021).

Nopal Cactus serves as the primary source for Desserto leather, known for its biodegradability, softness, and durability. With its organic and cruelty-free composition, this material finds versatile applications in fashion items, home furnishings, and car interiors. Cactus cultivation offers significant carbon dioxide absorption, making Desserto a promising eco-friendly choice for various leather goods (Williams & Trejo, 2024).

Figure 3.9 depicts the process flowchart of cactus leather production.

3.5 Apple Leather

Apple leather is made from the apple waste and the apple skin (Ashley, 2021). Initially, the waste is transformed into a powder before being combined with PU and coated onto a cotton/polyester canvas. This innovative blend consists of 50% apple waste and 50% PU, resulting in a durable yet soft fabric suitable for various accessories (Oliver—Apple Leather, 2024).

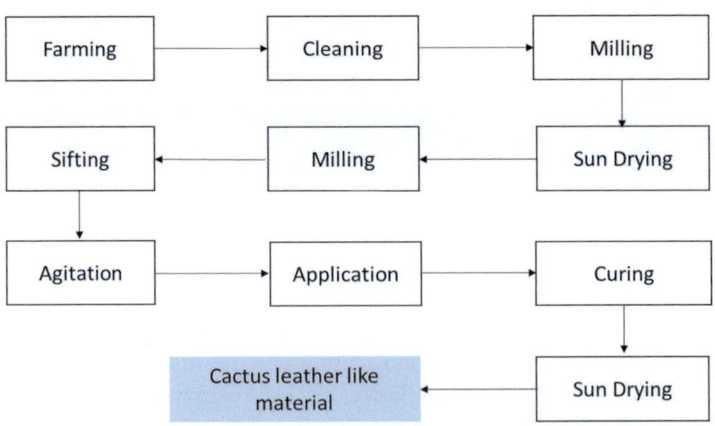

Fig. 3.9 Cactus leather production. (Source: Pettersen & Krych, 2023)

3.6 Cork Leather

The oak bark is harvested, dried, boiled, and pressed into sheets before being bonded to fabric backing using natural suberin adhesive. The resulting cork leather is a flexible, soft, and durable alternative, offering an environmentally friendly option for vegan leather products (Ashley, 2021; Mahi Leather—Cork Leather, 2024).

3.7 Grape Leather

Wine leather production involves repurposing waste materials generated by the wine industry, such as leftover grape remnants. These materials are combined with water-based PU and applied onto recycled polyester woven textile, resulting in a material resembling leather (Mahi Leather—Cork Leather, 2024).

3.8 Corn Leather

Corn leather is made by coating the corn-based polymer in a cotton fabric (Ashley, 2021).

3.9 Banana Leather

Banana leather production involves harvesting banana trees for their fibers, which are then processed, treated, and formed into sheets (Banofi Leather, 2024).

3.10 Soyabean Leather

Soybean waste is converted into microbial cellulose by boiling it with sugar, fertilizer, vinegar, and bacteria for 10 days. This cellulose is dried and processed into sustainable fabric, suitable for vegan leather production. Wastewater from tofu industries is used to culture Acetobacter xylinum bacteria, which convert glucose into cellulosic fibers over 10–12 days. These fibers can be further processed into microbial cellulose sheets for leather production. While formaldehyde is sometimes used in fabric production, soy fabrics remain a more sustainable option compared to artificial fibers (Gupta & Dave, 2021).

3.11 Bioleather1

Modern Meadow (MM) has introduced Bioleather1 (BL1), a cutting-edge class of next-generation materials. BL1 harnesses MM's proprietary Bio-Alloy™ technology platform, which exploits molecular-level interactions between protein molecules and bio-based polymers. This innovative approach empowers BL1 with superior material performance characteristics, making it ideal for demanding applications like footwear and accessories. By utilizing the Bio-Alloy™ technology, BL1 achieves exceptional durability and functionality while remaining environmentally conscious.

This innovative technology yields a high-performing material that is both animal-free and bio-based, capable of large-scale production using existing infrastructure. BL1 utilizes a Bio-Alloy™ composed of a miscible blend between plant-derived protein molecules and bio-based PU. This unique blend is created through a reverse transfer process, leveraging manufacturing lines typically used for conventional synthetic leather production. The plant protein utilized in BL1's Bio-Alloy™ is sourced from plants also producing oil for biofuel or food, as well as soymeal predominantly used in animal feed. Consequently, environmental impacts associated with farming and upstream operations must be distributed among all relevant products in these systems (Locker & Theregowda, 2022; Modern Meadow—Biomaterials, 2024).

3.12 Coconut Leather

Coconut leather originates from the fermentation of coconut water, commonly known as "Nata de coco," using Acetobacter xylinum bacteria. During fermentation, a jelly-like substance forms, which is sterilized and processed with natural fibers like pineapple, hemp, and banana. The resulting cellulose jelly is shaped into sheets and combined with banana fiber, known for its strength and lustrous texture. While thicker sheets offer durability, thinner ones are easier to handle and can be molded into various shapes using a natural blend of wax and oils. Coconut leather is robust, biodegradable, and environmentally friendly, gradually decomposing without emitting harmful gases or harming the soil (Gupta & Dave, 2021). Figure 3.10 clearly explains the production flow of coconut leather.

3.13 Orange Silk Leather

Utilizing waste citrus fruit peels, orange peel fiber is extracted for cellulose production, treated with sodium sulfite or sodium metabisulfite as squashing reagents. Process parameters such as the amount of sulfite agent and reaction period significantly affect cellulose yield. The resulting cellulose exhibits high purity, low

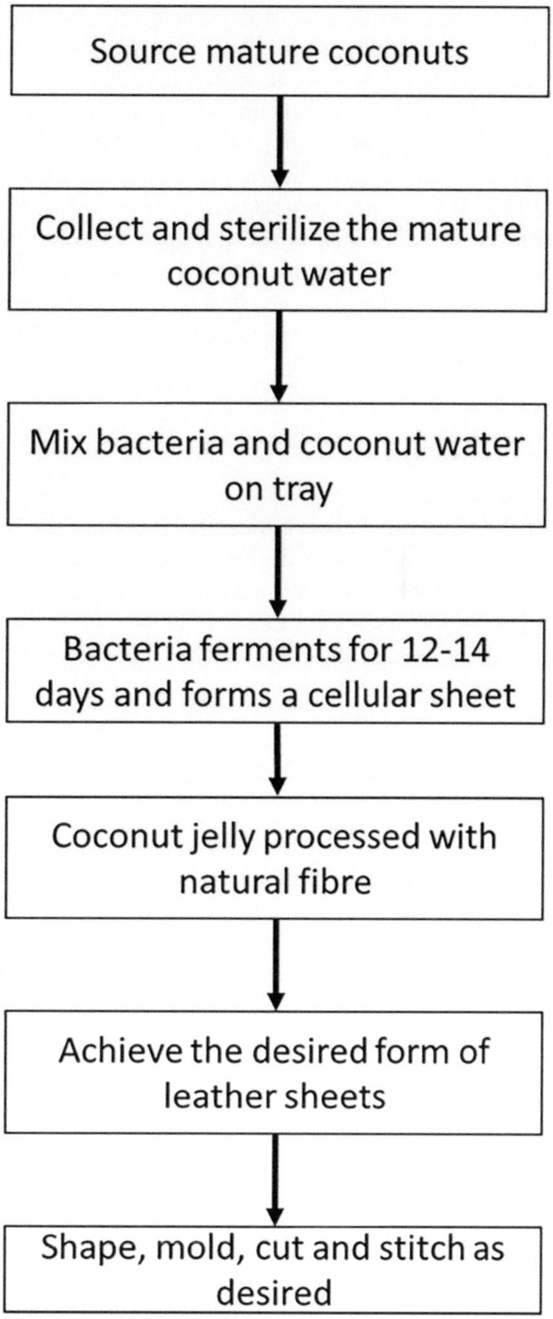

Fig. 3.10 Coconut leather production flowchart. (Source: Pettersen & Krych, 2023)

crystallinity, brightness, water retention, and reasonable molecular weights after bleaching with oxygen and hypochlorite. These cellulose fibers are then spun into yarns, resulting in a fabric with a luxurious appearance and silky texture. Beyond aesthetics, the fabric contains vitamin C and essential oils inherent in citrus fruit peels, providing nourishing properties to the skin without feeling greasy or uncomfortable. Additionally, the fabric retains these properties through multiple washing cycles. Orange peel extracts are also used for antimicrobial finishing on textiles, offering benefits such as anti-odor and antimicrobial properties, making them suitable for medical textiles. These fabrics, derived from waste materials, are soft, absorbent, and breathable, offering various treatment options such as fragrance finishes and mosquito repellents (Gupta & Dave, 2021).

3.14 Leaf Leather

Leaf leather production intricately layers sustainably harvested teak leaves with fabric, preserving nature's designs without harming animals. The resulting material is strong, durable, and water-resistant, ideal for various gear and accessories (Ashley, 2021; Tree Tribe—Leaf Leather, 2024).

The product made of leaf leather is shown in Fig. 3.11.

3.15 Mango Leather

Mango leather, a plant-based alternative to traditional leather, is crafted by mechanically separating the fruit's flesh from the seed and pulping it. Following this, binding agents are incorporated, and the mixture is baked in an oven. Leather finishers apply coatings to sheets of material removed from the oven. The mango leather, placed over an organic cotton base, displays satisfactory mechanical characteristics (Kefale et al., 2023).

Fig. 3.11 Leaf leather-like material. (Source: Tree Tribe—Leaf Leather, 2024)

3.16 Hemp Leather

Hemp fibers are harvested, cleaned, and treated to enhance pliability. They are combined with binding agents, molded, and finished with dyeing and coating processes. The result is a durable, eco-friendly alternative to traditional leather, offering both performance and sustainability benefits (Future Material Banks—Hemp, 2024).

3.16.1 Animal-Based Leather

The collagenous matrix found in animal skins, renowned for its durability, has been adapted for various purposes. Recent innovations aim to create collagen matrix alternatives without animal products, such as laboratory-cultured bioleather pioneered by Modern Meadow. By multiplying cow cells in a lab and nurturing collagen creation on a suitable medium, this technique produces a unique collagen structure suitable for bioleather (Jakab et al., 2019). The laboratory-cultured animal leather is shown in Fig. 3.12.

4 Life Cycle Assessment: Natural Versus Artificial-Derived Vegan Alternatives

The conventional animal leather and vegan leather impact for 1 m^2 is tabulated in Table 3.1 (Pettersen & Krych, 2023).

The climate change impact for kg CO_2 equivalence of bovine leather and Reishi leather (fungus-based) is represented in graph form in Fig. 3.13.

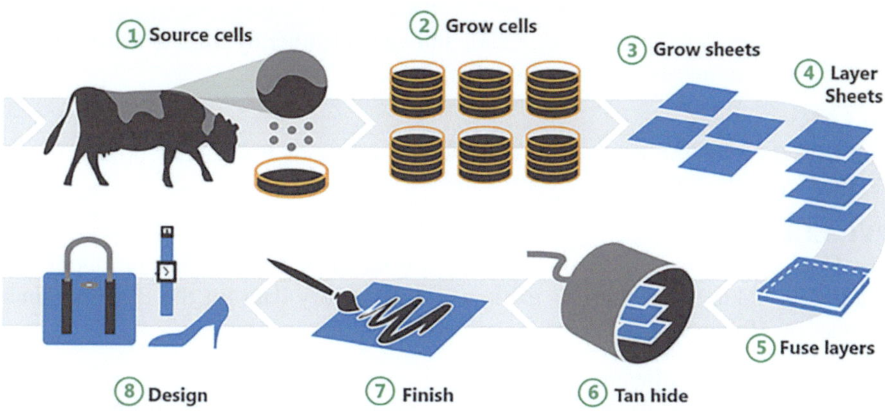

Fig. 3.12 Production flow of laboratory-cultured animal leather. (Source: Jakab et al., 2019)

Table 3.1 Environmental impacts per square meter for various leather types

S. no	Leather types	GWP (kg CO_2-eq/ m^2)	BWC (liters/ m^2)	PED (MJ/ m^2)
1	Bovine leather—Chrome-tanned	111.9	16495.3	131.37
2	Bovine leather—Vegetable-tanned	109.3	16,492	126.2
4	PU leather	3.9	268.05	97.93
5	PVC leather	2.93	30.21	77.78
6	Pineapple leather	2.69	144	107.46
7	Cactus leather	1.1	18.8	36.23
8	KBC (kombucha bacterial cellulose) leather	12.27	268.5	112.42
9	Mycelium leather	15.25	0.95	79.46

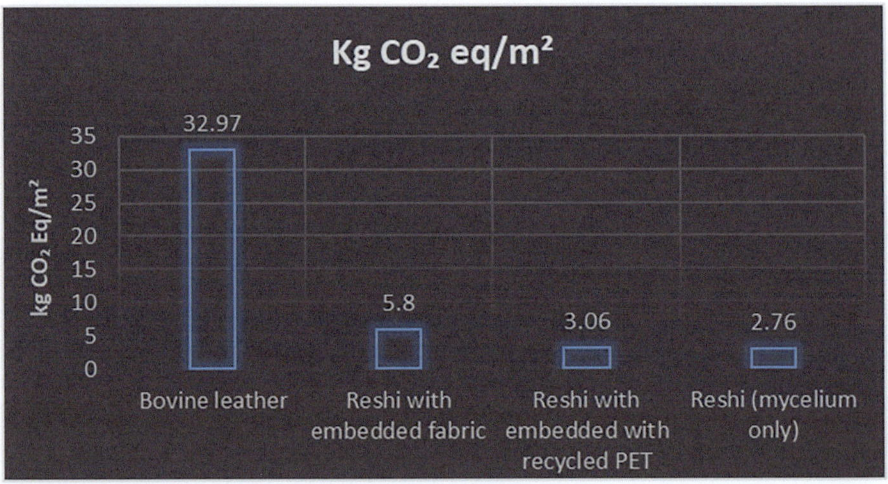

Fig. 3.13 Carbon footprint of Reishi. (Source: Jakab et al., 2019)

The comparison of traditional leather and vegan leather by Mother Earth depicts the clear impact numbers as shown in Fig. 3.14.

Pu LCI
The LCI data for the production of 1 square meter of PU leather production is detailed in Table 3.2.

PVC LCI
The LCI data for the production of 1 square meter of PVC leather production is detailed in Table 3.4. Table 3.3 explains the inventory data for the PVC polymer production (Table 3.4).

Pineapple Leaf Leather
The LCI data for the production of 1 square meter of pineapple leaf leather production is detailed in Table 3.5.

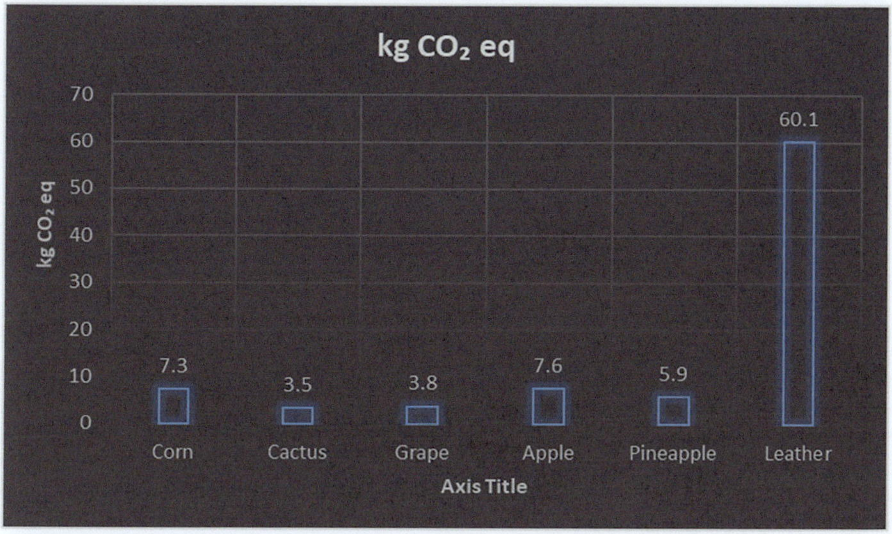

Fig. 3.14 Global warming potential (GWP) of traditional leather and different types of vegan leather. (Source: Mother Earth—Grape Leather, 2024)

Table 3.2 LCI to produce 1 m^2 of PU leather (Source: Pettersen & Krych, 2023)

Process	Activity	Value	Unit
Polymer production	Polyester polyol	0.23	kg
	Colorant	0.016	kg
	1,4-butyleneglycol	0.01	kg
	Dimethylformamide (DMF)	0.22	kg
	Diphenylmethane-4,4-diisocyanate	0.044	kg
	Electricity (homogenizer for 15 min + heating to 100 °C for 1.5 h)	0.945	kWh
Coating	Polyblend fabric (1 m^2)	0.15	kg
Submersion	Dimethylformamide (DMF)	0.13	kg
	Water (30 °C)	3.26	kg
	Heat (for 5 min)	0.15	MJ
Washing	Water (55 °C)	3.26	kg
	Heat (for 40 min)	0.94	kg
Drying	Electricity (drying at 100 °C for 20 min)	1.23	kWh
Output	1 m^2 of polyurethane leather-like material	0.326	kg

Cactus Leather

The LCI data for the production of 1 square meter of cactus leather production is detailed in Table 3.6.

Table 3.4 LCI to produce 1 m^2 of PVC leather production (Source: Pettersen & Krych, 2023)

S. no	Process	Activity	Value	Unit
1	First pour	Polymer solution	0.01	kg
		Silicone-coated paper (1 m^2)	0.07	kg
		Electricity (heated calender)	0.18	kWh
2	Second pour	Polymer solution	0.1	kg
		Riser (2% w/w)	0.002	kg
		Electricity (homogenizer for 5 min + heated calender)	0.428	kWh
3	Backing	Polyester fabric, nonwoven	0.15	kg
		Vinyl acetate (glue)	0.5	kg
		Electricity (heated calender)	0.178	kWh
4	Finishing	Acrylic varnish (1 wt.)	0.0033	kg
		Electricity (heated calender)	0.178	kwh
5	Output	1 m^2 of polyvinylchloride (PVC) leather	0.326	kg

Table 3.3 Polymer solution preparation for PVC leather (Source: Pettersen & Krych, 2023)

Process	Value	Unit
Input		
Polyvinylchloride (PVC) (polymer)	0.11	kg
Nonionic surfactant (plasticizer)	0.08	kg
Stabilizer	0.002	kg
Polysiloxane (filler)	0.004	kg
Anionic surfactant (filler)	0.006	kg
Electricity (homogenizer for 5 min)	0.25	kWh
Output		
Polymer solution	0.2	kg

Kombucha BC Leather

LCI data of BC production is given in Table 3.7. Table 3.8 details the LCI of natural dye production for the kombucha leather production. The LCI data for the production of 1 square meter of kombucha BC leather production is detailed in Table 3.9.

Impact Assessment of Hemp Leather

The functional unit is 100 m^2 of finished hemp leather sheet, with a thickness of 1.5 mm. The LCI data of hemp leather is detailed in Table 3.10.

5 Assessment Validity and Certifications

There are multiple organizations issuing certificates to label materials as "vegan". Vegan certificates for labeling materials as "vegan" are given below.

Table 3.5 LCI to produce 1 m² of pineapple leaf leather production (Source: Pettersen & Krych, 2023)

S. no	Process	Activity	Value	Unit
1	Decorticating	Pineapple leaves	9.6	kg
		Electricity (operation for 0.8 min)	0.1	kWh
2	Chemical degumming	Sodium hydroxide 2% (in a 50% solution)	9.6	kg
		Distilled water (95 °C)	230.4	kg
		Distilled water (22 °C, rinsing)	9.6	kg
		Heat (80 °C for 30 min)	87.48	MJ
3	Airlaying	Electricity (operation for 0.15 min)	0.02	kWh
4	Fusing	Polylactic acid (PLA)	0.081	kg
		Electricity (press at 120 °C for 30 sec)	0.22	kWh
5	Curing	Polyvinylchloride (PVC)	0.04	kg
		Electricity (press at 120 °C for 30 sec)	0.22	kWh
6	Calendering	Acrylic resin (finishing product)	0.004	kg
		Electricity (operation for 1 min)	0.447	kWh
7	Drying	Electricity (drying at 160 °C for 1 min)	0.062	kWh
8	Tumbling	PALF leather-like material	0.404	kg

Table 3.6 LCI to produce 1 m² of cactus leather production (Source: Pettersen & Krych, 2023)

S. no	Process	Activity	Value	Unit
1	Agriculture	Opuntia ficus-indica cladodes	1.62	kg
		Water	289.3	kg
		Manure	0.252	kg
		Potassium nitrate	0.003	kg
		Monoammonium phosphate	0.0005	kg
		Calcium nitrate	0.001	kg
		Potassium sulphate	0.014	kg
		Diesel	0.019	kg
2	Cleaning	Water	2.1	kg
		Sodium hypochlorite 15% (final concentration: 2%)	0.28	kg
3	Milling (twice)	Electricity (milling machine for 0.25 min)	0.057	kWh
4	Sifting	Electricity	0.001	kWh
5	Coating (four times)	Liquid polyurethane	0.476	Kg
		Electricity (homogenizer for 2.5 min)	0.25	kWh
6	Curing (four times)	Electricity (hot press at 150 °C for 1 min)	1.73	kWh
7	Sublimation	Polyester fabric, nonwoven	0.15	kg
		Electricity (sublimation machine for 10 min)	0.17	kWh
8	Output	1 m² of leather-like material from cactus	0.595	kg

OEKO-TEX

Leather material and leather goods, including footwear, are eligible for testing and certification at various points during processing. Leather meeting this standard

Table 3.7 LCI of kombucha BC leather production (Source: Pettersen & Krych, 2023)

Process	Activity	Value	Unit
Kombucha production	Green tea leaves	0.72	kg
	Symbiotic culture of bacteria and yeast (SCOBY)	7.2	kg
	Citric acid	0.08	kg
	Sucrose	3.6	kg
	Mineral water (22 °C)	72	kg
	Electricity (bioreactor at 30 °C for 14 days)	14.8	kWh
Output	Kombucha	69	kg
	BC (wet weight)	3.55	kg

Table 3.8 LCI of natural dye production (Source: Pettersen & Krych, 2023)

Process	Activity	Value	Unit
Natural dye production	Deionized water	3.3	kg
	Ethanol (70%)	1.11	kg
	Eucalyptus globulus (dry leaves)	0.22	kg
	Allium cepa L. (bulb bark)	0.22	kg
	Punica granatum (dried fruit)	0.22	kg
	Electricity (boiling for 30 min)	0.75	kWh
Output	Natural dyes	4.4	kg

Table 3.9 LCI of 1 m^2 of kombucha BC leather-like production (Source: Pettersen & Krych, 2023)

Process	Activity	Value	Unit
Cellulose purification	BC biofilm	3.55	kg
	Sodium hydroxide 0.1 M	0.043	kg
	Deionized water (22 °C)	10	kg
	Water (70 °C)	10.64	kg
	Electricity (water heating for 1 h)	1.48	kg
Pre-dyeing	Potassium alum (99.5%)	0.21	kg
	Water (90 °C)	10.64	kg
	Electricity (water heating for 30 min + agitator operation)	1.95	kg
Dyeing	Natural dyes	44	kg
	Water (22 °C, 90 °C)	20.64	kg
	Sodium chloride	0.11	kg
	Electricity (water heating for 1 h agitator operation)	28	kg
Cellulose reconstitution	Electricity (industrial blender for 5 min)	0.12	kg
Waterproofing	Wax from *Copernicia prunifera*	0.05	kg
	Oil from *Melaleuca alternifolia*	0.05	kg
	Electricity (wax homogenization at 75 °C for 15 min)	0.77	kg
Reinforcement	Polyester fabric, nonwoven	0.15	kg
	Vinyl acetate (glue)	0.05	kg
Output	Kombucha BC leather-like material	0.23	kg

Table 3.10 LCI of 100 m² of hemp leather (Source: Hultkrantz, 2018)

Inputs	Unit	Amount needed for manufacturing process	Amount needed, cultivation of hemp
Raw hemp fiber	kg	40	40
Electrical power	MJ	401	30.4
Water	m^3	15.3	0.32
Diesel	kg	4.64	0.17
Sodium bicarbonate	kg	0.94	
Ammonia gas	kg	0.14	
Sodium chromate (sodium dichromate)	kg	4.72	
Sulfuric acid	kg	3.22	
Cyclohexyl carboxyl diphenyl amino thiazolyl sulfonamide (cyclohexylamine)	kg	1.73	
Cresol	kg	74	
SBR resin	kg	121.8	
Ammonium nitrate	kg		0.2
Triple superphosphate	kg		0.1
Potassium chloride	kg		0.3
Calcium oxide	kg		0.89
Seed for sowing	kg		0.15
Outputs			
Hemp leather sheet	m^2	100	100
Wastewater and sludge	kg	11,540	0
Hazardous waste	kg	109.6	0
Other waste	kg	466	0

undergoes rigorous scientific testing to ensure the absence of harmful substances (OEKO-TEX® Leather Standard, 2024).

BeVeg
BeVeg is accredited under ISO 17065 as a published vegan standard and is considered the world's most trusted vegan certifier, recognized as the best vegan trademark for consumers and retailers globally (BeVeg, 2023). The certification logo is shown below in Fig. 3.15.

The Vegan Society
The Vegan Trademark is an internationally recognized vegan product certification, established in 1990 by The Vegan Society, and has been instrumental in making vegan products more accessible. Registration with the trademark gives brands the confidence to highlight their vegan credentials (The Vegan Society, 2024). The certification logo is shown below in Fig. 3.16.

V-Label
V-Label is an internationally recognized registered seal for labeling vegan and vegetarian products and services, established in Switzerland in 1996. It serves as a

Fig. 3.15 BeVeg logo. (Source: BeVeg, 2023)

Fig. 3.16 The Vegan Society logo. (Source: The Vegan Society, 2024)

reliable shopping guide for consumers (V-Label, 2023). The certification logo is shown below in Fig. 3.17.

Certified Vegan
Distributed and recognized globally, the Certified Vegan Logo is a registered trademark for products that do not contain animal products or by-products and that have not been tested on animals. The certified logo is easily visible to consumers interested in vegan products and assists vegans in shopping without constantly checking ingredient lists (Certified Vegan, 2023). The certification logo is shown below in Fig. 3.18.

6 Threats or Risks in Waste Disposal

Waste disposal in vegan leather production, like any industrial process, poses several potential threats and risks to the environment and human health. Here are some of the key concerns:

1. *Chemical Pollution.*
   The chemicals used in the production of vegan leather, such as synthetic polymers and various treatment agents, can pose risks if not properly managed during disposal. These chemicals may leach into soil and water sources, contaminating ecosystems and potentially harming aquatic life and agricultural lands.
2. *Microplastic Pollution.*
   Many vegan leather alternatives are made from synthetic materials like PU and PVC, which can shed microplastics during use and disposal. These microplastics can enter waterways and the food chain, posing risks to wildlife and human health.

Fig. 3.17 V-Label logo. (Source: V-Label, 2023)

Fig. 3.18 Certified Vegan Logo. (Source: Certified Vegan, 2023)

3. *Energy Consumption.*
   The production of vegan leather typically requires significant energy input, particularly for the synthesis of synthetic materials. Improper disposal of materials can result in wasted energy resources and contribute to overall environmental degradation.
4. *Landfill Overflow.*
   If not properly managed, waste from vegan leather production can contribute to landfill overflow. This can lead to increased methane emissions, groundwater contamination, and habitat destruction.
5. *Toxic By-products.*
   The production process of some vegan leather alternatives may generate toxic by-products, such as VOCs or heavy metals. Improper disposal of these by-products can result in soil and water contamination, posing risks to ecosystems and human health.
6. *Greenhouse Gas Emissions.*
   The disposal of vegan leather waste, particularly if incinerated, can contribute to greenhouse gas emissions, further exacerbating climate change.

7 Conclusion

Many such vegan alternatives require footprint information from their cradle to gate stages, at least initially, through LCA. Including consumer use and disposal phases to obtain final life cycle impact assessments is of paramount importance.

To mitigate environmental risks and impacts, it is crucial for companies in the vegan leather industry to prioritize sustainable practices throughout the entire lifecycle of their products, including waste disposal. This may involve implementing recycling programs, developing biodegradable alternatives, improving production efficiency to reduce waste generation, and ensuring proper treatment of chemical by-products. Additionally, regulatory oversight and industry standards can help enforce responsible waste management practices.

References

Anam Palf. (2024). Accessed from https://www.ananas-anam.com/. Accessed on 13 Apr 2024.

Ashley, V. (2021). Knowledge of leather alternatives: An exploratory study: Implications for education. *Journal of Textile Science & Fashion Technolog, 7*(4). https://doi.org/10.33552/JTSFT.2021.07.000668

Banofi Leather. (2024). Accessed from https://www.banofileather.com/. Accessed on 13 Apr 2024.

BeVeg. (2023). Accessed from https://www.beveg.com/. Accessed on 13 Apr 2023.

Biomaterials. (2024). Accessed from https://www.modernmeadow.com/. Accessed on 13 Apr 2024.

Bolt Technology—Mylo. (2024). Accessed from https://boltthreads.com/. Accessed on 13 Apr 2024.

Certified Vegan. (2023). Accessed from https://vegan.org. Accessed on 13 Apr 2023.

Desserto. (2024). Accessed from https://www.desserto.com/. Accessed on 13 Apr 2024.

Ekegårdh, N., & Kristersson, E. (2021). *From Fungi to Vegan Leather—A case study made to highlight the potentials of mycelium-based leather in the textile industry and the unique features of its value chain, 2021*. Accessed from https://www.diva-portal.org/smash/get/diva2:1654482/FULLTEXT01.pdf. Accessed on 13 Apr 2024.

Elkhateeb, W. A., Galappaththi, M. C. A., Wariss, H. M., Haesendonck, K. V., & Daba, G. M. (2022). Fungi-derived leather (mushroom leather). *MycoKing, 1*, 1–5.

Future Material Banks—Hemp. (2024). Accessed from https://www.futurematerialsbank.com/. Accessed on 13 Apr 2024.

Gupta, R., & Dave, D. (2021). Biomaterial: A sustainable alternative to animal leather and synthetic material. *Annals Of The Romanian Society For Cell Biology, 25*(6), 7317–7331.

Gurera, D., & Bhushan, B. (2018). Fabrication of bioinspired Superliquiphobic synthetic leather with self-cleaning and low adhesion. *Colloids and Surfaces A: Physicochemical and Engineering Aspects*. https://doi.org/10.1016/j.colsurfa.2018.02.052

He, S., Wang, Y., Li, X., & Liu, H. (2018). Research on recycling of waste leather produced in leather manufacturing process based on supply chain management. *Leather and Footwear Journal, 18*. https://doi.org/10.24264/lfj.18.2.1

Hultkrantz, M. (2018). An overview on the environmental impacts of synthetic leather made of hemp fiber with preliminary life cycle assessment. Degree project technology, First cycle, 15 credits, Kth royal institute of Technology, School of Architecture and the Built Environment, Stockholm, Sweden.

Jakab, K., Marga, F., Kaesser, R., Chuang, T. H., Varadaraju, H., Cassingham, D., & Forgacs, G. (2019). Non-medical applications of tissue engineering: Biofabrication of a leather-like material. *Materials Today Sustainability, 5*, 100018. https://doi.org/10.1016/j.mtsust.2019.100018

Kefale, G. Y., Kebede, Z. T., & Birlie, A. A. (2023). A systematic review on potential bio leather substitute for natural leather. *Journal of Engineering, 2023*(1), 1629174. https://doi.org/10.1155/2023/1629174

Locker, C. R., & Theregowda, R. (2022). Life-cycle assessment of Bioleather1. *Cleaner and Circular Bioeconomy, 1*, 100003. https://doi.org/10.1016/j.clcb.2022.100003

Mahi Leather—Cork Leather. (2024). Accessed from https://mahileather.com/. Accessed on 13 Apr 2024.

Mother Earth—Grape Leather. (2024). Accessed from https://www.moea.com/. Accessed on 13 Apr 2024.

Mushroom Leather—Watson & Wolfe. (2024). Accessed from https://www.watsonwolfe.com/. Accessed on 13 Apr 2024.

Nguyen, H. T., Saha, N., Ngwabebhoh, F. A., Zandraa, O., Saha, T., & Saha, P. (2021). Kombucha-derived bacterial cellulose from diverse wastes: A prudent leather alternative. *Cellulose, 28*, 9335–9353. https://doi.org/10.1007/s10570-021-04100-5

OEKO-TEX® Leather Standard. (2024). Accessed from https://www.oeko-tex.com/en/our-standards/oeko-tex-leather-standard. Accessed on 13 Apr 2024.

Oliver—Apple Leather. (2024). Accessed from https://www.olivercompanylondon.com/. Accessed on 13 Apr 2024.

Pettersen, J. B., & Krych, K. (2023). *Environmental performance of leather and leather-like materials applied to footwear.* Master's thesis. NTNU 2023. Accessed on 21 Apr 2024.

Reishi. (2024). Accessed from https://www.mycoworks.com/. Accessed on 13 Apr 2024.

Reuters. (2024). *How food and agriculture contribute to climate change*. Accessed from https://www.reuters.com/business/environment/factbox-how-food-agriculture-contribute-climate-change-2023-12-02/. Accessed on 13 Apr 2024.

The Vegan Society. (2024). Accessed from https://www.vegansociety.com/. Accessed on 13 Apr 2024.

Tree Tribe—Leaf Leather. (2024). Accessed from https://www.treetribe.com/. Accessed on 13 Apr 2024.

V-Label. (2023). Accessed from https://www.v-label.com. Accessed on 13 Apr 2023.

Williams, S., & Trejo, H. (2024). *Sustainable leather alternatives: A comparison of cactus leather mechanical properties*. Accessed from https://scholarworks.calstate.edu/downloads/jm214v95j. Accessed on 13 Apr 2024.

Williams, E., Cenian, K., Golsteijn, L., et al. (2023). Correction: Life cycle assessment of MycoWorks' Reishi™: The first low-carbon and biodegradable alternative leather. *Environmental Sciences Europe, 35*, 38. https://doi.org/10.1186/s12302-023-00741-4

Chapter 4
Natural/Agro-derived Versus Artificial Vegan Leather: How Leather Alternatives Influence the Sustainable Luxury and Fashion Industry

1 Introduction

The German language term “ersatz” goods means substitute or replacement good (Wikipedia, 2023a, b, c; Merriam-Webster, 2023). Specifically, in terms of quality, the ersatz good is of inferior quality when compared to the original. The ersatz goods were mainly produced during the wartime period in Germany, as the original product or raw materials were rationed. In the current scenario of the luxury and fashion industry, several brands are experimenting with sustainable alternatives for leather. For example, Stella McCartney and Adidas have used mycelium-based vegan leather Mylo™ for their limited-edition products. In the luxury watch sector, IWC Schaffhausen uses TimberTex, a paper-based material, for making watch straps (Gründer, n.d.) and Mirum®, a natural plant-based vegan leather. Desserto®, a cactus-based vegan leather from the company Adriano Di Marti of Mexico, has collaborated with H&M, Karl Lagerfeld, Mercedes-Benz, etc. (Desserto, 2023). An important question arising here is what motivates luxury and fashion brands to use vegan alternatives for leather. To some extent, the search and adoption of alternative materials that look and feel like leather emerge from the influx of reports on environmental pollution and degradation caused by the menace of the fashion industry. Additionally, there is a growing consumer awareness and demand for vegan leather products inculcating ethical values and animal welfare.

According to the research by Meyer et al. (2021), traditional leather over the years has been processed and adjusted with different properties depending on the applications. For example, leather for shoes is made hydrophobic, leather for gloves is soft, and hard leather texture is used for the sole of the shoe. Industrialization led to different alternative materials being invented, like coating textiles with linseed oil and natural rubber through a vulcanization process to remove stickiness. Next came

---

Mukta Ramchandani is the primary author of this chapter.

S. S. Muthu, M. Ramchandani, *Vegan Alternatives for Leather*,
Sustainable Textiles: Production, Processing, Manufacturing & Chemistry,
https://doi.org/10.1007/978-3-031-65365-0_4

the invention of synthetic polymers, which led to the replacement of leather for high-performance technical and design-based solutions like gear, conveyor belts, upholstery, shoes, etc. (Meyer et al., 2021).

To be a sustainable brand or a company inculcating sustainable practices, the right material search has taken precedence in recent years. The environmental impact from the fashion industry, like water pollution and energy emissions, has usually been considered toward the end of a product's life cycle. But there is a shift in this trend as companies are beginning to work with sustainable materials right from the start of the production process (Ali & Sarwar, 2010). This search within the leather industry has led to the development of materials that are naturally/agro-derived or artificially created and include plants like cactus, pineapple skin, apple, corn, grapes, or fungi like mushrooms, bacterial cellulose, etc.

1.1 How Luxury Is Different from the Fashion Industry

In the fast fashion industry, the primary focus is on mass production, faster time periods between collections, price affordability, trends, and fads (Ozdamar-Ertekin & Atik, 2015). Within the luxury industry, the following attributes are observed (Kapferer & Bastien, 2009):

- Limited production quantity to maintain exclusivity and product scarcity.
- High quality.
- High level of details and craftsmanship.
- Expensive raw materials and products.
- A long history or heritage of the brand.

In this context, there are also several brands that are premium and share similar traits of a luxury brand/product but are in a medium price range, e.g., Hugo Boss, Michael Kors, Coach, etc. The emphasis in both these sectors (luxury and premium goods) is on aesthetics, quality, and exclusivity for the consumers, which are also some of the factors determining the willingness of luxury consumers to pay more.

In the leather goods industry for luxury and fashion consumers, the distinction also must be made based on their hedonic and utilitarian motivations to buy a product. A utilitarian motivation would fulfill a consumer's functional need, for example, when buying footwear for the purpose of walking/running, while the hedonic motivation aims at satisfying the desirability of a product and having an emotional connection, for example, buying a limited-edition Christian Louboutin shoe with the red-colored sole. Along these attributes, newer materials are manufactured by the "market-driving" luxury brands/companies in the name of innovation. Sustainability may not be a primary objective for manufacturers and suppliers when introducing newer materials, especially in the context of leather goods, where design and profit orientation currently take precedence (Ramchandani & Coste-Maniere, 2020).

1.2 The Need for Sustainability in the Luxury and Fashion Industry

The luxury industry and products by essence are meant for durability and are argued to be more long-term sustainable (Coste-Manière et al., 2015). But due to the faster production times and lack of sincerity in maintaining a long-term commitment to sustainable raw material sourcing, the luxury industry is scrutinized for its environmental impact. For example, the sourcing of animal-based materials (leather and fur) lacks transparency for consumers in terms of origin, labor rights, and animal welfare. Buyers and consumers of luxury and fashion products are increasingly becoming aware of sustainability issues and the environmental disclosure from brands. Tools like the Fashion Transparency Index (Fashion Revolution, 2023) rank retailers and brands on their social, environmental, and supply chain impact, contributing to the sustainability initiatives within the industry and the transparency disclosures from luxury and fashion brands (Jestratijevic et al., 2018). Moreover, consumers within the luxury and fashion industry are seeking brands that are sustainable. However, consumers could often overlook the materials used in the products and are satisfied with the brand's marketing and communication, consequently increasing the probability of being greenwashed by the brands in the name of sustainability. For example, many consumers of vegan leather products are unaware that it could be made with PVC and PU (Ramchandani and Coste-Maniere, 2020). A common trend by fashion companies is also the use of synthetic materials derived from recycled PET (polyethylene terephthalate) bottles and being marketed as an eco-friendly alternative to leather for bags, clothes, or shoes. But companies do not communicate about the shedding of microplastics in the environment from these new materials (Bryce, 2021). Moreover, the high demand for these materials (like the recycled PET bottle fabric) is surpassing the supply and availability.

2 Current Scenario and Trends

The fast fashion industry as well as the luxury industry uses traditional leather for producing garments, shoes, bags, furniture, watches, etc. With market demands for finding sustainable alternatives to leather, several companies have developed leather look-alike materials that are marketed as cruelty-free and environmentally friendly. Is there a shift from traditional leather to vegan leather in the name of sustainability? In this context, several challenges are observed and discussed.

2.1 Challenges

2.1.1 Organoleptic

Referring to the touch, feel, smell, and vision aspects of a product, organoleptic evaluation is also an important consideration when developing new alternatives for leather in the luxury and fashion industry. At the conscious and subconscious level, sensorial attributes of a product can influence consumer judgment and decision-making. Some common findings understood from the vegan leather suppliers in the industry are that consumers are mainly looking for alternatives that "look and feel" like the texture of genuine leather and are satisfied with their purchase as a cruelty-free product, meaning not derived from animals. This sensorial aspect is crucial to maintain for the luxury brands to satisfy the hedonistic motivation in the adoption of vegan leather. Some consumers may also be allergic to genuine leather, so they can consider using hypoallergenic vegan leather made with sustainable dyes.

2.1.2 Mechanical Properties

The prospect of substituting animal leather with an artificial one must also satisfy the mechanical properties such as the following (Spinnewijn, 2022):

- *Tensile strength*: To understand how strong and rigid a material is, tensile strength is evaluated.
- *Tear strength*: It is performed to assess the resistance of a material against tear.
- *Flex resistance*: The durability of a material can be assessed with the flex resistance, and its levels indicate the probability of cracks in the material.

For example, apple skin-based artificial leather is used in making belts in Italy, but the clients are unaware that the shelf life of such a product is only 2 years and may not be long-term durable (Belts manufacturer, personal communication MIPEL Milano, 2022).

2.1.3 Product Versatility

The opponents of vegan leather equate multi-usage with industry-wide application. Indeed, historically genuine leather has found multifaceted applications across footwear, bags, belts, saddlery, clothes, furniture, etc. But it does not mean that there must be a single perfect vegan alternative fulfilling the traditional conformities. A particular vegan alternative, e.g., cork leather, does not stand out to be a high-performing material in terms of longevity for bags, but when used inside the shoe lining and sole, it is a robust material. To develop an artificial leather alternative may not always be versatile or universal in application. For example, Mirum® as a plant-based leather alternative is used for fashion, upholstery, footwear, and accessories but is not used for making belts due to the size limitation (Belts manufacturer, personal communication Lineapelle Milano, 2023).

2.1.4 Sustainable Production

Certain types of vegan leather may be dependent on their natural resources for production or the geographical conditions. For example, Piñatex, a pineapple-based vegan leather, comes from the Philippines, and cactus-based vegan leather Desserto® comes from Mexico. This may be a geographically limiting factor for supplying to the fast fashion companies involving bulk production. On the other hand, for the luxury industry, it could be the right choice to maintain product scarcity, exclusivity, and control over the supply. The product circularity aspects of the 4R—reduce, reuse, recycle, and recover—are being addressed by some of the brands active in the vegan leather industry, evidence of which can be seen by the LCA reports on their websites (e.g., Desserto, Mirum). These are instrumental indicators of a product's traditional circularity approach, but newer extensions like refuse, rethink, reduce, reuse, repair, refurbish, remanufacture, repurpose, recycle, and recover must also be incorporated for a product's design, clean production, and regenerative design in sustainability (Kircherrer et al., 2017; Hildebrandt et al., 2021; Reike et al., 2017). Consequently, sustainable production by essence could coincide with eco-design to holistically be part of the product circularity in the luxury and fashion industry.

3 Natural/Agro-based Leather Alternatives

One of the first artificially created imitations of leather was produced around the Second World War in Germany, due to the shortage and rationing of raw materials needed for leather. The new imitation leather was called *Presstoff* and made of layered and treated paper pulp (Wikipedia, 2023a, b, c). This imitation leather was able to substitute for leather applications like equipment belts, binocular cases, and straps, but it was unsuitable for footwear due to delamination and losing cohesion easily under regular wear and tear or moisture conditions. Natural/agro-based leather alternatives are materials derived as a substrate from natural materials, agricultural waste/residue during production, bacterial/microbial cellulose, or mycelium composites (Hildebrandt et al., 2021). The aim of some vegan leather materials is to biomimic genuine leather attributes; therefore, some companies use chitin, glucans, and proteins derived from fungal mycelium cultivation (Feofilova, 2010). Historical industrial usage can be traced back to the 1950s, when fungal biomass was used to create fabrics and conventional papermaking (Amobonye et al., 2023).

In an interview with Mr. Velarde, Co-Founder of Desserto®, a vegan leather company from Mexico, he explained that there is a difference between "plant-based" and "bio-based" vegan leather materials. As a material can be 100% bio-based, but one must consider the percentage extracted from plants. In the case of Desserto, it is 90% plant-based, and depending on the customer's demand and the application, the company can produce different variations (Velarde, personal communication, 2023).

Currently, the synthetic leather made with PU and PVC satisfies the mechanical properties moderately when compared with genuine leather. However, the main issues highlighted with synthetic vegan leather are that these are plastic-based non-biodegradable materials. The PVC production itself uses chemicals increasing the toxicity levels of the material by releasing dioxins (Cao et al., 2013; Spinnewijn, 2022), and PU production uses the chemical N,N-dimethylformamide (Zhao et al., 2018; Spinnewijn, 2022), which is harmful to the environment and marine life. Therefore, companies are making efforts to produce bio-based and bio-sourced PVC and PU to mitigate the environmental impact.

The following are some examples of vegan leather:

3.1 Desserto®

Founded by the company Adriano Di Marti in Mexico (Desserto, 2023), Desserto is a type of vegan leather made with cactus (nopal: Opuntia ficus-indica)-based biomaterial, with applications in the fashion, automotive, and furniture industries. The cactus plant itself is a perennial plant lasting for about 8 years and is resilient, growing in semiarid regions with restricted water availability. It helps relieve the accumulation of CO_2 in the atmosphere and does not compete with the food industry (Desserto, 2023). The company also owns an organic ranch in Zacatecas, Mexico, for growing cactus raw material and uses rainwater for irrigation. No herbicides or pesticides are used at the farm, and residual cactus materials not used in Desserto production are sold to the food industry.

The process involves the following (Desserto, 2023):

- Harvesting mature cactus leaves every 6–8 months without damaging the whole plant.
- Rinsing and shredding into thin fibers.
- Drying for 3 days under the sun (no additional energy is used) to create a leather-like material.
- Mixing with nontoxic chemicals to add color and shape.

Figures 4.1, 4.2, 4.3, 4.4, and 4.5 illustrate these steps.

It has biodegradable properties and uses recycled cotton in some of its linings and backing materials. It is free from toxic chemicals and has a lower carbon and water footprint. Following the standards of ISO 14040 and 14,044 for LCA guidelines (Desserto, 2023), the materials rank better in sustainability than its synthetic alternative PU and traditional animal leather.

Fig. 4.1 Shredded cactus

Fig. 4.2 Cactus farm

Fig. 4.3 Cactus harvesting

Fig. 4.4 Post-shredding and drying

Fig. 4.5 Final product cactus leather

3.2 *Fleather*

Fleather is a type of vegan leather material made by fermenting discarded flowers from temple wastes in India (Vegan First, n.d.; Phool, n.d.; Pathak, 2022). It tackles the problem of wasted flowers from temples and mosques in India and uses them in a fermenting process to obtain a SCOBY-like material resembling the texture of leather. Currently, Fleather is used for making wallets, bags, and accessories that are biodegradable and free from plastics.

3.3 *Malai*

Malai is a vegan leather brand making leather-like material from coconuts and discarded coconut water. It is currently used to make bags, wallets, and pouches (Malai, 2023).

3.4 *Mirum®*

Mirum is made with a combination of plant-based natural materials like natural rubber, agro-based fillers (coconut husk fiber, cork powder), plant-based oil, pigments, and minerals (NFW, 2023a, b). Made in Illinois, USA, this vegan material has a texture like genuine leather and is used for making fashion accessories, bags, wallets, etc. Some brands using Mirum are Stella McCartney, IWC Schaffhausen, Camper, etc.

3.5 *Piñatex*

Piñatex (Ananas Anam, 2023) is a pineapple-based vegan leather from a London-based company Ananas Anam. The company produces vegan leather by using pineapple leaves in the Philippines, which are then combined with PLA (polylactic acid), "bioplastic" from cornstarch, and a petroleum-based resin (Wikipedia, 2023a, b, c). It is used in making bags, shoes, wallets, etc., and brands like Puma and Camper have used it in their footwear collections.

3.6 *Hanji*

Hanji is a type of paper-like material made using mulberry leaves, native to Korea and other parts of Asia. It is not a vegan leather by essence, as its texture does not resemble genuine leather. However, there are variations of Hanji paper that feel like the texture of "suede leather." From the perspective of substituting leather and finding an alternative, it fulfills criteria such as circularity, mechanical strength, durability, and biodegradability. The bark of mulberry trees contains lignin and holocellulose (Wikipedia, 2023a, b, c), contributing to Hanji's durability. Historically, the material has been used for making armors, clothes, textiles, and arts and crafts known as "jigap" in Korean. Korean artist Ms. Kim Hyunjoo of KHJ Studio makes wallets, pouches, and luxury art pieces from Hanji. These products are sold and showcased in museums and available to retailers worldwide (KHJ Studio, 2023).

3.7 *Other Examples*

Pliant™ is used in making outsoles in the footwear industry (NFW Pliant, 2023a, b). Reishi is a fine mycelium vegan leather developed by MycoWorks (Reishi, 2023). Vegatex produces AppleSkin, LemonSkin, and BarleySkin made from apple pomace, lemon pomace, and barley, respectively, which are combined with a water-based PU and lyocell for backing material (Appleskin, 2023). VEGEA® makes vegan leather from grape leftovers from the winemaking industry in Italy (VEGEA, 2023).

4 Conclusion

It is yet to be assessed what the long-term implications are of vegan leather being used by the luxury and fashion industry and how consumers respond to the latest innovations. Rendering benefits of vegan leather and substituting genuine leather come with their own challenges as discussed earlier in this chapter. Developing the texture and odor of leather artificially is possible with the use of synthetics or a combinatory approach with natural materials (plant-based or bio-based). Further research can investigate if vegan leather companies can drive the market in the luxury and fashion industry, and how luxury brands using vegan leather can retain their value and desirability for sustainable impact.

References

Ali, M. A., & Sarawar, M. I. (2010). *Sustainable and environmental friendly fibers in textile fashion a study of organic cotton and bamboo fibers*. Master's Thesis. University of Borås, Borås, Swedish School of Textiles. Retrieved December 18, 2023, from https://www.diva-portal.org/smash/record.jsf?pid=diva2%3A1312071&dswid=6211.

Amobonye A., Lalung, J., Awasthi, M.K, Pillai, S. (2023). Fungal mycelium as leather alternative: A sustainable biogenic material for the fashion industry, Sustainable Materials and Technologies, Volume 38, https://doi.org/10.1016/j.susmat.2023.e00724.

Ananas Anam. (2023). Retrieved December 5, 2023, from https://www.ananas-anam.com/

AppleSkin. (2023). Retrieved December 10, 2023, from https://www.appleskin.com/

Belts manufacturer, Personal communication, Lineapelle Milano Italy 2023. https://www.lineapelle-fair.it/en

Belts manufacturer, Personal communication, MIPEL Milano Italy, 2022. https://mipel.com/en/

Bryce, E (2021). Retrieved December 12, 2023, from https://www.theguardian.com/environment/2021/nov/06/clothes-made-from-recycled-materials-sustainable-plastic-climate).

Cao, H., Wool, R., Sidoriak, E., & Dan, Q. (2013). Evaluating mechanical properties of environmentally friendly leather substitute (eco-leather). In: *International textile and apparel association annual conference proceedings* (Vol. 70, No. 1). https://www.iastatedigitalpress.com/itaa/article/id/2042/.

Coste-Manière, I., Ramchandani, M., Sudeep, C., & Burak, C.(2015). "Long-Term Sustainable Sustainability in Luxury. Where Else?" In Handbook of Sustainable Luxury Textiles and Fashion, volume 2. Springer Science+Business Media, Singapore, 2016.

Fashion Revolution. (2023). Retrieved December 18, 2023, from https://www.fashionrevolution.org/about/transparency/

Feofilova, E. P. (2010). The fungal cell wall: Modern concepts of its composition and biological function. *Microbiology, 79*, 711–720. https://doi.org/10.1134/S0026261710060019

Gründer, E. (n.d.). Retrieved December 18, 2023, from https://www.iwc.com/en/articles/journal/paper-straps.html.

Hildebrandt, J., Thrän, D., & Bezama, A. (2021). The circularity of potential bio-textile production routes: Comparing life cycle impacts of bio-based materials used within the manufacturing of selected leather substitutes. *Journal of Cleaner Production, 287*. https://doi.org/10.1016/j.jclepro.2020.125470

Jestratijevic, I., Uanhoro, J., & Rudd, N. A. (2018). Policies versus practices: Transparency of supply chain disclosures among luxury and mass market fashion brands. In *International textile and apparel association annual conference proceedings* (Vol. 75, No. 1). Iowa State University Digital Press.

Kapferer, J. N., & Bastien, V. (2009). *The luxury strategy- break the rules of marketing to build luxury brands*. Kogan Page Publishers.

KHJ Studio. (2023). https://www.khjstudio.com/

Kirchherr, J., Reike, D., & Hekkert, M. (2017). Conceptualizing the circular economy: An analysis of 114 definitions. *Resources, Conservation & Recycling, 127*, 221–232.

Made with Reishi. (2023). Retrieved December 10, 2023, from https://www.madewithreishi.com/products

Malai. (2023). Retrieved December 10, 2023, from https://malai.eco/blogs/malai-material/malai-applications

Merriam-Webster. (2023). Retrieved December 18, 2023, from http://www.merriam-webster.com/dictionary/ersatz

Meyer, M., Dietrich, S., Schulz, H., & Mondschein, A. (2021). Comparison of the technical performance of leather, artificial leather, and trendy alternatives. *Coatings, 11*, 226. https://doi.org/10.3390/coatings11020226

NFW. (2023a). Retrieved December 10, 2023, from https://www.nfw.earth/materials-mirum

NFW. (2023b). Retrieved December 10, 2023, from https://www.naturalfiberwelding.com/pliant

Ozdamar-Ertekin, Z., & Atik, D. (2015). Sustainable markets: Motivating factors, barriers, and remedies for mobilization of slow fashion. *Journal of Macromarketing, 35*(1), 53–69. https://doi.org/10.1177/0276146714535932

Pathak, S. (2022). https://www.bbc.com/future/article/20221208-fleather-the-vegan-leather-made-of-indias-flower-waste

Phool. (n.d). https://phool.co/pages/our-story

Ramchandani, M., & Coste-Maniere, I. (2020). Leather in the age of sustainability: A norm or merely a cherry on top? In S. Muthu (Ed.), *Leather and footwear sustainability. Textile science and clothing technology*. Springer. https://doi.org/10.1007/978-981-15-6296-9_2

Reike, D., Vermeulen, W. J., & Witjes, S. (2017). The circular economy: New or refurbished as CE 3.0?—Exploring controversies in the conceptualization of the circular economy through a focus on history and resource value retention options. *Resources, Conservation and Recycling*.

Spinnewijn, O. (2022). *Innovative leather alternatives: A short- lived trend or the future of fashion?* https://studenttheses.uu.nl/handle/20.500.12932/41443

Vegan First. (n.d). https://www.veganfirst.com/article/fleather-the-new-vegan-and-eco-friendly-leather-solving-gangas-floral-wasteproblem

VEGEA Company. (2023). Retrieved December 10, 2023, from https://www.vegeacompany.com/

Velarde, A. (2023). Personal communication interview of the Co-Founder Desserto.

Wikipedia. (2023a). Retrieved December 18, 2023, from https://en.wikipedia.org/wiki/Ersatz_good#cite_note-2

Wikipedia. (2023b). Retrieved December 18, 2023, from https://en.wikipedia.org/wiki/Presstoff

Wikipedia. (2023c). https://en.wikipedia.org/wiki/Pi%C3%B1atex

Zhao, B., Qian, Y., Qian, X., Fan, J., Liu, F., & Duo, Y. (2018). Preparation and properties of split microfiber synthetic leather. *Journal of Engineered Fibers and Fabrics, 13*(2). https://doi.org/10.1177/155892501801300203

Chapter 5
Producers' Dilemma: Craftsmanship, Credibility, and Pricing Substrates of Vegan Versus Traditional Animal-Based Leather

1 Introduction

Traditional animal-based leather has found its applications across several industries, including the luxury and fashion industry, where it is seen as a versatile raw material for making shoes, bags, belts, upholstery, saddlery, art, furniture, clothes, etc. It is praised as a robust material and has brought, throughout history, sufficient profits and margins for the producers who supply to the market demand. But these margins are slowly disappearing with increases in costs of production, governmental norms in pollution and wastewater treatment, reduction in the supplier base, and hesitation of smaller factories in adopting innovative sustainable methods in leather processing and tanning. Additionally, consumers are becoming more aware of the environmental costs that come with traditional leather production. In this regard, with the onset of newer developments, it becomes imperative to understand how attractive it is for manufacturers to produce vegan alternatives for leather. What motivates these producers to manufacture vegan leather? And how do brands and designers in the luxury and fashion industry perceive the new vegan leather materials? The aim of the current chapter is to understand these issues and elucidate what further challenges persist for both producers and designers.

The industry and factory setups of traditional animal-based leather have had their roots primarily in factors such as accessibility to animal hides and skins, tanning materials, and technical skillsets. In the pre-industrial economy, leather was seen as a scarce material, and its supply was dependent on a country's cattle assets and the structure of the meat industry. For instance, animal hides and skins were defined as "residual products" of the meat industry in France and England, where the price of leather was dependent on the number of cattle, turnover, and the amount of meat consumed in these countries (Riello, 2008).

Mukta Ramchandani is the primary author of this chapter.

S. S. Muthu, M. Ramchandani, *Vegan Alternatives for Leather*,
Sustainable Textiles: Production, Processing, Manufacturing & Chemistry,
https://doi.org/10.1007/978-3-031-65365-0_5

Tanning agents derived from vegetable substances (certain barks from trees and fruits) and mineral oils have also contributed to the scarcity of leather and the perception around it. For instance, the British government during the mid- and late-nineteenth-century set up leather factories in India to meet the leather demand for boots and shoes for its army. These tanneries were set up in regions like Kanpur, Kolkata, Agra, Chennai, and others (Bhargava, 2022) as the traditional vegetable tanning agents derived from babul (Acacia arabica), myrobalan (*Terminalia chebula*), and avaram bark (Cassia auriculata) were easily accessible (Sharma, 1946). Moreover, the setup of these leather factories in different regions led to changes at the organizational level, demand for new products and training the workforce in these regions, and diffusion of new techniques and technology in leather tanning and processing (Bhargava, 2022), except for chrome tanning, which came much later to the Indian tanneries (in the mid-1930s) when compared with its counterpart tanneries in Europe. This shift from vegetable tanning to chrome tanning led to the change in accessibility of leather due to faster and more efficient production.

However, with the rampant use of traditional leather globally, the industrial production became unsustainable over the years, especially with the chemical-based tanning agents and the increase in cattle rearing for the leather industry instead of it being the by-product of the traditional meat industry. Other issues have emerged besides animal slaughtering for the leather industry. For example, illegal deforestation in Brazil, Uruguay, Argentina, and Paraguay has been linked to their animal ranches required for leather production and the leather export industry (Earthsight, 2020; De Koning, 2020), which also raises the issue of traceability in the supply chain of companies importing leather worldwide.

Certainly, it has become clear over the years that the traditional leather industry falls short of making amendments to be able to fulfill the industrial demand for leather and at the same time enforcing and implementing methods and processes to reduce its impact on the environment. Hence, research and development of artificial leather look-alike materials performing like leather have paved the way as a suitable replacement for traditional leather, not just in its application but also in its adoption by brands and consumers worldwide. But limited investigation is done on understanding to what extent these vegan alternatives are profitable and what factors the production companies are dependent on for their setup.

Through both primary and secondary data collected, the current chapter aims to shed light on these issues mentioned above. Three interviews were conducted for this book to uncover and provide depth on this topic. The interviewees include a vegan leather company's co-founder, a shoe designer, and an artist. The details of the interviews and methodology are described in earlier chapters of this book.

2 Vegan Leather Production and Challenges

Production and manufacturing of traditional leather, as described earlier in this chapter, have had their own roots and factors which determined their survival. Contrastingly, vegan leather alternatives do not rely on these same factors as animal slaughtering, but, having distinguished themselves with plant-based, bio-based, agro-derived, fungi-derived, etc., they are subsequently also dependent on certain raw materials found in nature. To assess this, it is important to first understand where the motivation for producers and designers begins in the search for vegan leather materials.

Vegan leather company Adriano Di Marti produces leather alternatives from cactus (Desserto® and Dessertex®) (see Fig. 5.1) and has found its motivation for production from several factors. According to Mr. Velarde (interview, personal communication, 2023), he and his co-founder have had previous experience within the textile, fashion, automotive, and furniture industries where they had a closer technical assessment of the different materials used in the industry and the negative environmental impact caused by materials like leather. Moreover, the availability of raw materials in Mexico was limited at that time to leather and synthetic materials, which were also imported from other countries. The idea for them was to be able to go around these roadblocks, and they began their research into materials which could also be made in Mexico. Consequently, this led them to evaluate cactus, which is an endemic plant with inherent sustainability advantages over other plants and is symbolic of Mexico.

Fig. 5.1 Desserto cactus-derived vegan alternative to leather

For Mrs. Min Zhou Carpentier, the owner and designer of the French footwear brand ÉMZI Paris, it is primarily the style and usage on which the materials are selected (interview, personal communication, 2023). This is followed by the availability of the materials in the region, as all the brand's footwear is made in Europe, especially in Spain and Italy. Her preference for using natural materials lies in the fact that the style she creates has functional and long-term usage. For example, two types of soles are used in her company: traditional animal-based leather and rubber. The rubber used is not 100% natural but a composite to make it more flexible, soft, shock-absorbent, and long-lasting for walking every day on the street. Most importantly, she focuses on the functionality and performance of the footwear for long-term usage by customers. See Figs. 5.2, 5.3, 5.4, and 5.5 for various types of footwear soles.

Sustainable natural materials are also used by the designer and artist Ms. Hyunjoo Kim of KHJ Studio in Seoul, South Korea (interview, personal communication, 2023). Her creations are inspired by nature, and she believes in using materials that are sustainable and long-lasting and can be biodegradable. For example, the pouches and small bags (see Figs. 5.6 and 5.7) she produces are made with Hanji paper, which is a type of paper-like material derived from the mulberry tree (as described previously in Chap. 4).

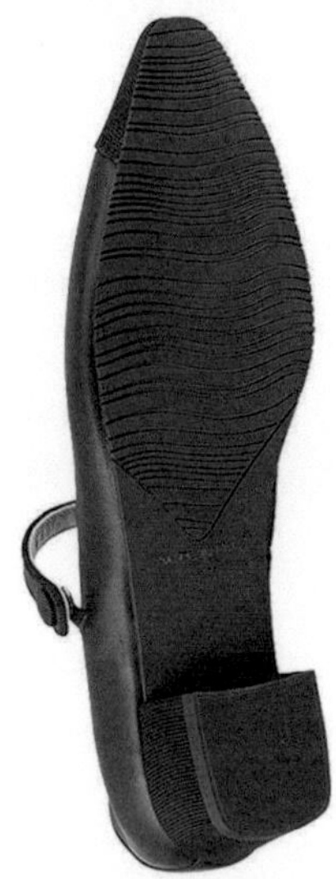

Fig. 5.2 ÉMZI light rubber sole

Fig. 5.3 ÉMZI extra-light micro sole

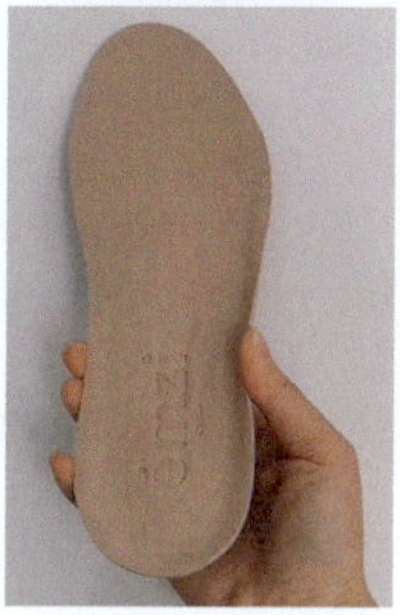

Fig. 5.4 ÉMZI footbed comfort insole

2.1 Technical Skillset: Skilled/Unskilled Workers

One of the goals of the vegan leather industry is to become a replacement for traditional animal-based leather. But without the knowledge and experience of agriculture and environment, these new vegan materials may be causing direct or indirect harm to the natural environment. Consequently, technical skillsets are needed to produce high-quality, solution-oriented vegan leather materials that perform like traditional leather and can also cover the specific demands of the market.

For example, in Desserto, a team of biochemists, industrial engineers, agricultural engineers, and experts work together not just at the laboratory level but also at the industrial level (Velarde, interview, personal communication, 2023). Their current production is standardized and efficient, producing 500,000 square meters per month in each of the two production points, showing readiness at the industrial level. This is also evidenced by their collaboration with fashion brands like H&M, Fossil, Adidas, and Onitsuka Tiger. Additionally, the company fosters a balance between innovation and costs along with its product customizations suited for

Fig. 5.5 ÉMZI leather sole

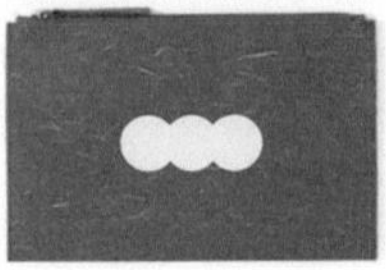

Fig. 5.6 Pouch from Hanji paper by KHJ Studio

different applications. For Desserto, alignment with agricultural practices is also part of sustainability, such as working with local cactus farms and uplifting agricultural communities through innovation and collaboration.

For other vegan leather materials, limited data is available on what kind of technical skillset and technologies are used by the producers, especially in the case of mycelium-based vegan leather—almost making it a secret. We tried contacting and interviewing a mycelium-based company in South Korea producing vegan leather for the automotive industry, but the company did not agree to be interviewed and share their perspective.

2.2 Impact on Factories and Production Units

While several vegan leather producers are innovating and developing products with agricultural wastes, plant-based, fungi-based, etc., and focusing on the biodegradability of these materials, current research is still limited in understanding the impact on factories and production units, especially for these materials to be scalable. For example, Mylo, one of the mycelium-based vegan leather production companies, has paused production because of scalability issues (Chan & Webb, 2023). Another issue is that the size and capacity to produce is limited for some vegan materials. For instance, the largest size of Fleather (the flower-based vegan leather) per sheet is 9 sq. ft. with a capacity to produce 9000 sq. ft. per month (Pathak, 2022). Fleather currently cannot be used for products like belts, as the tensile strength is between 6 and 10 megapascals, whereas for genuine leather, it is between 10 and 25 megapascals (Pathak, 2022). Another vegan leather brand, Malai, has a production capacity of 200 m. Their website states that it can be difficult to achieve exact Pantone® colors for them and there could be about 10% color variation when using 100% natural plant-based dyes (Malai, 2023). Most designers within the fashion

Fig. 5.7 Floral pouches from Hanji paper by KHJ Studio

industry use the Pantone® color codes, which not only facilitate clear communication between manufacturers, designers, and retailers but also provide precise color matching references. Therefore, if achieving precise colors on vegan leather materials is challenging, it can limit the growth of such materials in applications for the fashion industry.

Thus, the above-discussed factors could deter designers and brands from adapting newer vegan leather materials. However, this leads to the following questions: why is a vegan alternative to leather expected to be exactly the same as leather in terms of size, coloring, and universal applications? Why can't these materials be seen as a new innovation with limited applications customized to specific product type (depending on the material's unique attributes and availability) instead of being a "one-size-fits-all" approach?

2.3 *Economic Challenges: Following Money*

Within the animal-based leather and the fashion industry, it is common that economic incentives precede sustainable-driven changes. Without financial gain, these companies can be reluctant to adopt sustainable practices such as reducing chemical usage, conducting materials assessments, and obtaining product or process level certifications (Resta et al., 2014; Karaosman et al., 2020). Moreover, for sustainability-oriented innovation, positive collaborative relationships with supply chain actors and companies initiating change can lead to better outcomes than coercive methods (Brun & Cicculo, 2022). From this perspective, the adaptability of vegan leather by current suppliers in the fashion industry may still lag behind. Additionally, many of the current top luxury brands, especially French and Italian ones like Hermès, are deeply rooted in the leather industry. Completely switching to vegan leather could potentially disrupt existing relationships with suppliers and affect their current customer base. Therefore, establishing trust in newer vegan leather alternatives through supplier connections and relationships is crucial for brands as part of innovation and knowledge-sharing in the luxury and fashion industry.

2.4 Commercial Feasibility and Price

Industrial availability of vegan leather material means that it can be widely accessible for mass production to be used for products like footwear, bags, accessories, automotive, saddlery, etc. It also means that designers should be able to easily produce the products and have materials available to them or their manufacturing partners. This is why credibility between the producers of vegan leather, suppliers, and designers needs to be established. Generally, the MOQ (Minimum Order Quantity) must be met for a production agreement between the designer/brand and the factories producing the products. For example, a bag producer may source its leather from a tannery and a designer may be asked to have an MOQ of 300 bags per color per design.

It is also expected that the new materials are cheaper due to consumer demand for vegan materials to replace animal-based leather. This is evidenced through the influx of PU- and PVC-based fashion products in the market. However, the assessment of sustainability impact in the new vegan leather materials (not using PU or PVC) is still an ongoing process, and as such, research and development, especially in improving the performance of products and industrial scalability, could currently account for higher prices. For instance, Hanji paper is not very cheap compared to synthetic PU, but it is equally priced to good-quality animal-based leather and has a longer life compared to PU. On average, products made from Hanji paper by KHJ Studio, like handmade pouches, cost between 13 and 20 euros depending on the design (Hyunjoo Kim, interview, personal communication, 2023). On the other hand, Desserto's vegan leather is cheaper than animal-based leather and costs on average about 30 USD per linear meter or 21.5 USD per square meter for materials with 1.4 m width (Velarde, interview, personal communication, 2023).

2.5 Geographical Challenges

While newer plant-based or bio-based materials that resemble leather may be developed, their sourcing from different regions may also pose a challenge. For example, Piñatex is sourced from the Philippines and finished in Spain (Ananas Anam, 2024). Another issue with food-based or fruit-based vegan leather materials (like grapes, apples, oranges, etc.) is that they could directly conflict with the existing food industry in those regions. Traditionally, food waste from farming has been used to feed cattle. Sourcing a material from a distant region could also lead to higher CO_2 emissions and influence its sustainability impact. This argument applies similarly to the traditional animal-based leather industry, where leather is currently exported from different continents. This poses a similar challenge for the vegan leather industry, and the geographic sourcing of materials must be accounted for in the final product's sustainability impact.

Additionally, it may also be difficult for the industry to meet the demand for these materials, as access to suppliers is currently limited and, for some, there can be long waiting times. For instance, cork-based vegan leather, with its potential applications in footwear, bags, accessories, furniture, etc., is a sustainable material. However, it is primarily sourced from the regions of Portugal and Spain, where each cork oak tree can be harvested every 9–13 years, with the first harvest ready when the tree is about 25 years old (Wikipedia, n.d.).

3 Conclusion

The current chapter describes some of the challenges related to vegan leather production and usage by designers. As understood from the three interviews, there is a common factor of sustainable natural materials availability and its environmental impact which motivates producers of vegan leather, as seen in the case of Desserto and products designed by KHJ Studio using Hanji. For designers and brands in the footwear industry like ÉMZI Paris, style, usage, and natural materials procured from nearby regions count for the material selection in their collections, with a major focus on how these products perform for customers in everyday use.

Technical skillsets within the vegan leather industry can be manageable for certain companies with a strong base in research and development or access to agricultural resources (for plant-based vegan leather), but this may be challenging and a barrier for smaller companies or new suppliers entering the market. Therefore, in this regard, for industrial scalability of vegan leather, training and communication with interested designers or brands would be necessary, similar to the knowledge-sharing that has been crucial in the animal-based leather industry. Moreover, for those transitioning from traditional leather to new vegan alternatives, perception change is necessary. From these perspectives and challenges discussed in this chapter, vegan alternatives for leather should be viewed as completely innovative materials with applications similar to traditional leather. As exclusivity of materials is core to a product and its branding in the luxury industry, the limited production or scalability of vegan leather materials can enhance their status as "exotic" or "rare," potentially commanding higher prices. This can increase attraction and brand equity in the long term while also offering a sustainable and environmentally friendly alternative to animal-based leather.

References

Ananas Anam. (2024). Retrieved January 5, 2023, from https://www.ananas-anam.com/faqs/

Bhargava, P. (2022). Modernization of leather industry and chequered history of technical education in colonial Kanpur. *Indian Journal of History of Science, 57*, 102–114. https://doi.org/10.1007/s43539-022-00039-z

Brun, A., & Ciccullo, F. (2022). Factors affecting sustainability-oriented innovation in the leather supply chain. *Strategic Change, 31*(3), 305–321. https://doi.org/10.1002/jsc.2500

Chan, E., & Webb, B. (2023). Retrieved December 2, 2023, from https://www.voguebusiness.com/sustainability/stella-mccartney-backed-leather-alternative-mylo-halts-production-bolt-threads-kering-ganni-adidas-lululemon

De Koning, P. C (2020). *Deforestation related to beef & leather supply chains in Latin America and export to Europe*. Mekong Ecology. https://mekonecology.net/wp-content/uploads/2020/12/Mekon-Ecology-2020-Beef-Leather-Supply-Chain-Latin-America-Europe-vFINAL.pdf

Earthsight. (2020). *Grand Theft Chaco*. Retrieved December 2, 2023, from https://www.earthsight.org.uk/grandtheftchaco-en#group-Summary-R5FDaSdXCx

Karaosman, H., Perry, P., Brun, A., & Morales-Alonso, G. (2020). Behind the runway: Extending sustainability in luxury fashion supply chains. *Journal of Business Research, 117*, 652–663.

Malai. (2023). Retrieved December 14, 2023, from https://malai.eco/blogs/malai-material/material-range

Pathak, S. (2022). https://www.bbc.com/future/article/20221208-fleather-the-vegan-leather-made-of-indias-flowerwaste

Resta, B., Bandinelli, R., Ciarapica, F. E., Dotti, S., Pinto, R., & Rinaldi, R. (2014). Practices for environmental sustainability in the textile, clothing and leather sectors: The Italian case. *International Journal of Operations and Quality Management, 20*(3), 193–225.

Riello, G. (2008). Nature, production and regulation in eighteenth-century Britain and France: the case of the leather industry. *Historical Research, 81*(211), 75–99. https://doi.org/10.1111/j.1468-2281.2007.00413.x

Sharma, T. R. (1946). *Location of industries in India*. Hind Kitab Limited.

Wikipedia. (n.d.). Retrieved December 2, 2023, from https://en.wikipedia.org/wiki/Cork_(material)

Chapter 6
Reconciling the Sustainable Consumer Behavior Dimensions: The Lack Luster of Vegan Leather in the Luxury and Fashion Industry

1 Introduction

Many animals, especially exotic animals (e.g., alligators, crocodiles, snakes, etc.), are skinned for the purpose of deriving leather. In major parts of the world, these animals are not eaten and are mainly killed for the purpose of leather. The animal hides from cattle like buffaloes, cows, pigs, and goats are derived mainly as a by-product of the food industry, making it less prone to wastage as all the parts of the cattle are used. In some cultures, it is also considered disrespectful to waste parts of the animal when killed for food. This is one of the reasons humans historically found different uses for animal-derived raw materials (e.g., tallow, gelatin, adhesives, etc.). Traditional tenets of vegetarian and vegan consumption in some regions worldwide coincide with spirituality and religious beliefs. For example, in Hinduism, Buddhism, and Jainism, the concept of "Karma" and the principle of nonviolence are widely known. Hence, many people use vegan products in their everyday life and refrain from using animal-derived products like leather.

Within the luxury and fashion industry, there is a wide use of animal-derived materials like leather, fur, wool, cashmere, etc. Such materials share a reputation of being not just natural but also rare. Consequently, they fetch higher market prices based on their finer quality, which is liked by luxury consumers, adding to the exclusivity of these products. For example, the Italian luxury brand Loro Piana is known for developing and processing some of the finest cashmere, vicuña fiber, and merino wool sourced from different regions like China, Mongolia, the Andes, Australia, and New Zealand (LVMH, 2024). In the haute couture and luxury industry, brands also have a history of developing or sourcing these materials, accounting for strong brand equity for both buyers and investors. Additionally, this brand image of luxury brands resonates with the value of the product in terms of both prices and

---

Mukta Ramchandani is the primary author of this chapter.

S. S. Muthu, M. Ramchandani, *Vegan Alternatives for Leather*,
Sustainable Textiles: Production, Processing, Manufacturing & Chemistry,
https://doi.org/10.1007/978-3-031-65365-0_6

consumer's perceptions. For example, ZILLI, a French luxury brand known for its trademarked leather and exotic skins like ostrich, crocodile, kangaroo, etc., uses them for making limited-edition jackets, clothes, bags, and accessories (Global Times, 2014). Within the watch industry, many brands use watch straps made from leather or exotic skins, especially from crocodile, alligator, python, and other reptiles. However, with the increase in awareness about the inhumane slaughtering and skinning of these animals, certain regions are banning the sale of luxury products using exotic skins (e.g., California). Therefore, several luxury brands in both the luxury watch and fashion sectors have resorted to the following practices:

1.1 Responsible Sourcing and Traceability of Exotic Skins and Leather

For example, Swatch Group uses *Alligator mississippiensis* from sustainable farming in the USA (Chandrasekhar, 2020). Givenchy, Louis Vuitton, and Dior follow the LVMH group's own standard of responsible sourcing of crocodile skin (Castle, 2022). Kering invested in VitroLabs Inc., which bioengineers animal tissue in labs through stem cell technology and develops lab-based animal leather without animal slaughtering or harmful tanning processes (Vitro Labs Inc, 2024). Additionally, the Leather Working Group certification helps several brands in traceability and sustainable sourcing of leather (Leather Working Group, n.d.).

1.2 Banning the Use of Exotic Leather in Their Collections

For example: Burberry and Chanel.

1.3 Using Only Vegan Leather

For example, Greubel Forsey watch brand uses only vegan leather for its watch straps and permanently removes animal-based leather (SJX, 2021). Stella McCartney uses only vegan materials.

Research in the luxury industry (Kapferer, 2012) indicates that new materials like faux fur or fake leather are considered less precious and therefore do not add to the luxury appeal (dream factor) of a product. Furthermore, this results in a shift in the industry from objective rarity (initially a result of scarce, inaccessible materials like exotic skins) to qualitative rarity. Keech et al. (2020) found in their research that different message appeals for lab-grown diamonds and lab-grown leather have different impacts on luxury consumers. Their research indicates that an ethical positioning strategy is more effective, and the willingness to pay is higher among low materialist consumers compared to high materialist consumers, driven by the higher perceived social status among low materialist consumers. Since the high usage of leather and exotic skin in luxury products has formed an intricate part of luxury consumption, it raises the following questions: What role does consumer interest and motivation play in this industry? How does it position itself within the sustainable luxury industry? And how can the consumer perception be changed for vegan leather to make it more attractive?

2 The Role of Consumer Interest and Attitude in the Sustainable Luxury and Fashion Industry

Physical motivational drivers of a luxury product like the design, craftsmanship, quality, materials, and functionality are some of the reasons why consumers may purchase luxury products. Additionally, emotional and psychological drivers also play a motivational and attitudinal role in influencing consumer decision for a luxury product. For example, a luxury product could evoke desire, success, prestige, exclusivity/need for uniqueness, conspicuousness, or hedonism within consumers (Vigneron & Johnson, 2004; Dubois & Laurent, 1994; Leibenstein, 1950; Wiedmann et al., 2007). Integrating these could help assess different consumer groups globally and their value perception toward luxury products and brands. Moreover, cross-cultural assessments can also be made for different consumers in this context. For instance, owning a Mercedes-Benz car has a higher luxury appeal and brings status and prestige to consumers in Asian countries like India, China, and South Korea than in European countries like Germany and Switzerland. Similarly, Zara has a reputation for being a fast fashion brand in European countries but is considered a premium brand in Asia. In addition to value-based perceptions, there are other indicators like multiculturalism, individualism, and collectivism (Ramchandani & Coste-Maniere, 2012; Hofstede, 1980) which indicate different parameters for luxury consumption.

From the abovementioned perspectives, the current research is limited in assessing where vegan alternatives for leather stand within the sustainable luxury industry. Additionally, what type of consumer groups prefer vegan alternatives for leather? To assess these, it is necessary to distinguish different consumers and their goals (see Fig. 6.1). For example, a consumer may search for a vegan leather product which is as follows:

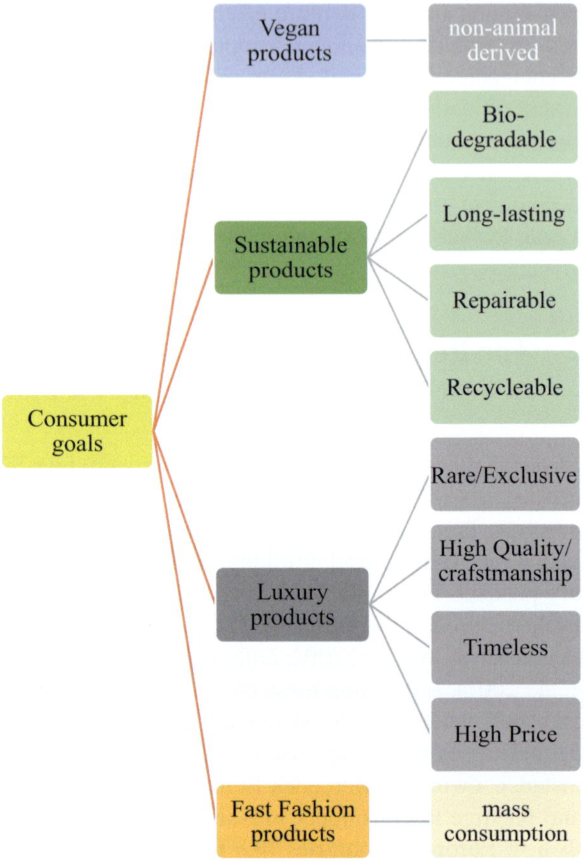

Fig. 6.1 Consumer goals

2.1 Sustainable

Meaning that the products are repairable, biodegradable, recyclable, or long-lasting.

2.2 Fast Fashion

Meant for utilitarian consumption and mass consumption.

2.3 Luxury

Material or product meant for hedonic consumption while retaining similar attributes of traditional animal leather, such as exclusivity, rarity, longevity, timelessness, high quality, and craftsmanship.

This does not mean that the three goals are always mutually exclusive. A combinatory approach of these can lead to the following scenarios where "bags" as a product type is taken as an example:

- A vegan bag may be sustainable if completely made with natural materials (leather look-alike or not), e.g., a jute bag, cotton canvas bag, cork, or Hanji bags.
- A vegan bag may be fashionable and meant for utilitarian purposes, e.g., a classic tote bag made with any vegan material, including natural or artificial alternatives like PU.
- A vegan bag may be luxurious, e.g., a limited-edition Hermès bag made with Fine Mycelium™ or a bag made with Mirum® which is limitedly produced.

2.4 *Vegan and Non-vegan Consumer Groups*

Previous research shows differences between vegan and non-vegan consumer groups and their perceptions toward luxury brands in adapting vegan alternatives. Park and Nunes (2024) investigated attitude toward luxury brands from vegan and non-vegan consumer groups. They found that non-vegan consumers have a higher positive perception (compared to vegan consumers) and higher brand trust, competence, and benevolence perception toward luxury brands adapting vegan alternatives. With the attribution theory as the theoretical background, they explained that vegan consumers are often doubtful about the real intentions of luxury brands and attribute their new measures in adapting vegan materials as a strategic decision rather than an ethical or moral one.

It is often confusing for brands which options are right for the market to launch and which is the right targeted audience for vegan leather because they think that both vegans and non-vegans would be their potential clients. While the non-vegan consumers are more open to vegan products, vegan consumers, on the other hand, deter from buying vegan leather products if they are produced by the same brands using animal-based leather in other collections. This is primarily because vegan consumers prefer brands selling only vegan product assortments as it is cruelty-free and the company as a whole does not engage in any form of animal suffering or animal testing (including products like cosmetics). Moreover, it is already clear from previous research (Kahn, 2011; Park & Nunes, 2024) that vegan consumers are skeptical of buying vegan products from brands selling animal-based products. Therefore, there is a need to separate the consumer groups and match their goals while assessing which vegan materials are suited for them. In this chapter, from a practical approach, we step further into distinguishing their needs of consuming leather and vegan alternatives "separately" without overlapping them and have identified which alternatives are available to them. For this purpose, the three goals (mentioned earlier in Fig. 6.1) are combined with the two different consumer groups (vegan and non-vegan) to provide further clarity in this book (Fig. 6.2).

2.4.1 Vegan Consumer

From the perspective of leather in the luxury and fashion industry, vegan consumers here can be described as those who refrain from using animal-based leather products. The primary motivation is using animal-free vegan products. Vegan alternative product-wise categories can be further divided as follows (see Fig. 6.2):

Sustainable

Full or semi-biodegradable natural/agro-derived/bio-based vegan alternatives, e.g., Fleather, Desserto, cork, Hanji, mycelium, etc.

Non-sustainable

Artificial vegan leather or petroleum-based, e.g., PU or PVC.

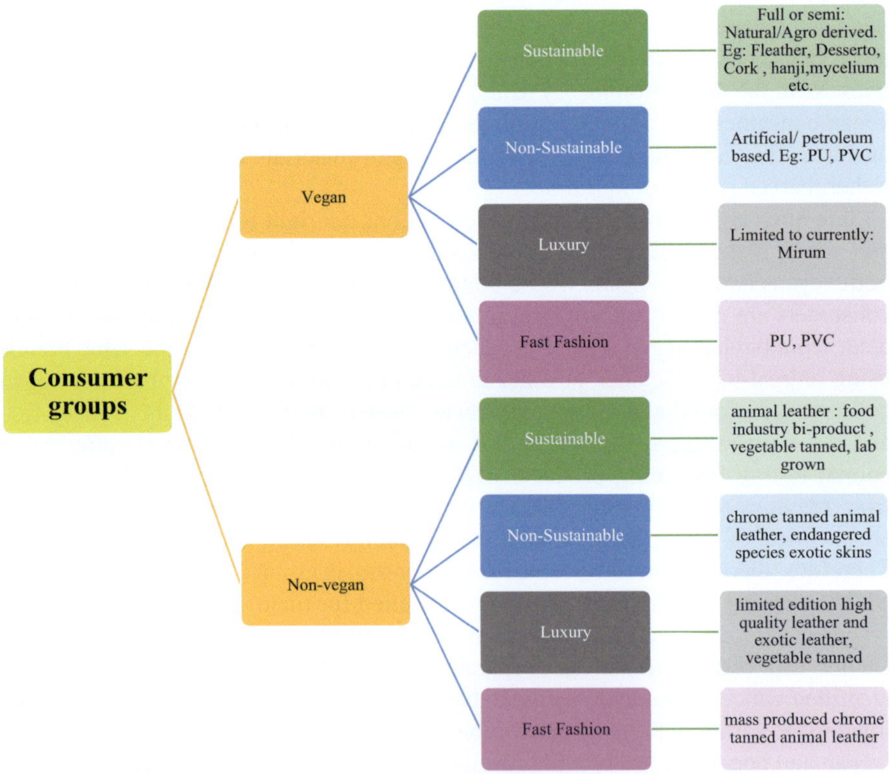

Fig. 6.2 Consumer groups and product-wise categorization

Luxury

For example: Mirum®.

Fashion

For example: PU or PVC.

This does not mean that all the four products are mutually exclusive; for example, sustainable and luxury makes a "sustainable luxury category," and Mirum® would be well-suited to this group as well.

2.4.2 Non-vegan Consumer

These are consumers who use animal-based leather/exotic skins. Primary motivation does not necessarily include animal-free products but rather set product-wise goals (see Fig. 6.2). This does not mean that non-vegans never use vegan products, but this group is differentiated here for clarity based on animal-derived product categories available to them:

Sustainable

Animal leather which can be sourced from the food industry as a by-product, can be vegetable-tanned following an environmentally friendly tanning approach, and can be lab-grown animal leather (e.g., VitroLabs Inc. making stem cell-based cow leather in the lab (Vitro Labs Inc, 2024).

Non-sustainable

Chrome-tanned animal leather, endangered species exotic skins like ostrich, crocodile, alligator, python, etc.

Luxury

Limited-edition high-quality responsibly sourced leather and exotic leather, vegetable-tanned.

Fashion

Mass-produced chrome-tanned animal leather.

Sustainable and luxury can also be a single group here, as described earlier for the vegan category. For non-vegan sustainable luxury category, it would also cater to high-quality limitedly produced vegetable-tanned cow leather.

3 Application Convergence but a Need for Divergence with Traditional Leather

Now that the different consumer groups and their motivations are illustrated with respect to the different options available to them in the leather and vegan leather industry, it is imperative also to understand where vegan leather converges with traditional animal-based leather beyond consumer perception and if there is a need for divergence in the market, especially for brands.

A major issue is that vegan alternatives are expected to be an exact look-alike of traditional animal-based leather, which, as described in previous chapters, is challenging due to different mechanical and organoleptic properties and other parameters. Simple sustainable vegan products are available in the market, like cotton canvas or jute, which can be used for making bags, clothes, belts, etc., but these are not considered luxury or desirable enough for luxury buyers. The utilization of different vegan leather alternatives as raw materials can be found in various products, and some of them have limited production capacities and scalability issues or fall short of real environmental impact. These applications are expected to have a long-term aim of replacing animal-based leather but also at the same time fight the menace of environmental degradation. Another issue is that vegan alternatives are all judged by most consumers as plastics-based, which is incorrect, as each type of vegan material has different properties and applications (described also in the previous chapters).

From the perspective of veganism and animal welfare, a true vegan product (alternative leather product) must consist of materials not derived from animals. It means that a complete product like footwear, if called vegan, must be made of only vegan materials from the upper, sole, lace guard, heel cap, lining, glue, etc. From the perspective of functionality, a vegan product must perform at its best ability for everyday use, and consumers must find their utilitarian needs fulfilled. While sustainability is a broader construct and includes animal welfare (e.g., vegan), environmental welfare (e.g., pollution-free, toxicity-free), and social welfare (e.g., fair wages, uplifting communities), within the scope discussed in this chapter, a sustainable product must be biodegradable, long-lasting, repairable, reusable, recyclable, upcycle, etc. From the perspective of luxury, a luxury product must be made of exclusive raw materials, high in price, and rare, include high level of craftsmanship, and fulfill a hedonic or aspirational need for consumers. However, considering the luxury watch industry, for example, a highly technical product from brands MB&F, Hublot, Richard Mille, or Patek Philippe would obviously fetch a higher value due to the level of craftsmanship involved in watchmaking, irrespective of whether animal-based or vegan leather is used in the watch straps, unless there is a shift in consumer perception where vegan leather itself is seen as a material of high craftsmanship.

4 Innovation and Finding a Place for Vegan Leather in the Luxury Industry

From the outlook of consumers, it is also likely that brands do not necessarily communicate or even understand the innovations that are proceeding in the vegan leather space. One reason could be that there is a shortage of people working within the luxury and fashion brands who understand clearly what sustainable vegan alternatives are available in the market and how they can innovate in developing them. Consumers no longer need to be marketed another new vegan alternative that is detrimental to the environment (like PU or PVC ones), merely for the sake of including a vegan capsule collection and marketing itself on this trend. Rather, sustainable vegan alternatives that are limitedly produced need to be highlighted and promoted for better adoption in the luxury market. Several certifications can also be adapted by luxury brands to be transparent to consumers, e.g., VEGANOK, Forest Stewardship Council (FSC), Global Recycled Standard (GRS), People for the Ethical Treatment of Animals (PETA)-approved vegan, etc.

The questions related to consumer behavior and perception of luxury were also asked in the three interviews collected for this book (methodology details described in previous chapters).

During the interview with Ms. Kim Hyunjoo of KHJ Studio in South Korea, it was noted that consumer perception is still strong for real leather as luxury. Despite the problems associated with pollution, especially water usage and wastewater treatment of leather, consumers still attach their perception of leather as luxury. However, they are willing to use alternative vegan materials like Hanji, which are less polluting and use less water.

In the footwear industry, Mrs. Min Zhou Carpentier of ÉMZI Paris stated that a perfect company does not exist yet. Brands using vegan materials face challenges such as some vegan leather easily cracking or being harder to repair and the PU in the sole can cause sweating and making it unbreathable. Currently, in her company, both animal-based leather and natural vegan alternatives like linen in some collections are used, all locally sourced from Spain, France, and Italy by road transportation, which has lower carbon emissions than air transport. The focus is on making limited-edition products using natural materials that are repairable and long-lasting. These, in addition to comfort, are of prime importance in her view so that consumers can wear the product for longer and repair them instead of discarding them, thus reducing waste. Mrs. Min Zhou Carpentier emphasizes that both brands and consumers must make conscious efforts to reduce environmental impact for sustainability.

The long-lasting feature of a product mentioned above by Mrs. Min Zhou Carpentier for footwear resonates with luxury attributes. In essence, luxury products can be passed on from one generation to another. For example, a leather bag from Hermès also fetches a high value in the secondhand marketplace. This attribute of passing vegan products from one generation to another while appreciating in value has not been found with vegan alternatives, especially the plant-based/bio-based ones (as per our current knowledge while writing this book).

Regarding consumer motivation in luxury to use vegan leather products, Desserto® co-founder Mr. Velarde stated in the interview that consumers like innovation and that luxury is a consequence of acceptance and desirability. He explained that as consumers need to feel part of the transformation in the industry, the Desserto material is perfect for making customizations and meeting the complicated needs of a designer or brand. He further added that consumers sometimes think something is luxurious because it is made of exotic skins like crocodile leather, a perception shaped by the marketing of materials by certain brands that use them. For many consumers, luxury means not wearing parts of dead animals or being part of this chain that grows crocodile skin, for example. Luxury for these consumers is wearing materials that align with their values, which are changing the industry and are also scarce. Something made of Desserto can be purchased from fewer brands, but it is also not mainstream yet. This is not because Desserto cannot produce it but because there is still a huge market size that is not yet using it. In his opinion, plant-based vegan materials represent only 0.1% of all materials circulating in the world. There is important work to be done between brands and innovative materials: to communicate, educate, and market the concept of luxury as something linked with sustainability.

5 Short-Term and Long-Term Approaches for Consumer Behavior in the Luxury Vegan Leather Industry

Consumer behavior research so far is aligned with finding values, attitudes, perceptions, motivational approaches, and other variables in marketing for purchasing vegan leather. Research by Keech et al. (2020) shows that materialism levels among consumers are an important indicator that luxury brands must assess to develop effective communication strategies for lab-made leather products. Moreover, luxury is perceived to be more valuable with "made in Italy or France" labels (Kumar & Gautam, 2021; Prin, 2000). However, from the perspective of sustainability, consumers should not always categorize these labels as inherently good or bad, as manufacturing products like footwear is a complex process that requires different expertise for different types of footwear. For example, the majority of Nike sports shoes are made in China but are of high quality and comfort. They can be challenging to produce in Italy or Spain, which are primarily known for leather footwear craftsmanship (Min Zhou Carpentier, interview, 2023).

As the options in general are differentiated for luxury vegan buyers, fashion vegan buyers, and sustainable (and non-sustainable) vegan buyers, as described previously in Fig. 6.2, there needs to be further consumer research conducted on these specific consumer groups and the options of vegan leather available to them, instead of overlapping all the options conducted for short-term forecasting. For instance, in the long-term approach, it could be researched what attitudinal and affective responses are generated from emotional appeals of different vegan leathers (luxury (Mirum) vs sustainable (cork) vs unsustainable (PU leather)). Further research can be conducted to investigate how luxury vegan leather is perceived by Generation Z vs Millennials vs Generation X vs Baby Boomers, as different age groups have different perceptions of sustainability and luxury (Michaut & Kapferer, 2020). Similarly, different cultural and country-wise consumer behavior can be tested in this context. The main goal here is not only to predict consumer adoption or behavior for just another line of innovative products like the new vegan materials, which can also be tested, for example, using Rogers' diffusion of innovation theory (Rogers, 1962), but the goal is rather long-term and threefold:

- How can a stronger correlation of vegan leather with luxury be strengthened?
- How can luxury brands succeed without compromising material sustainability and rarity/exclusivity of products?
- How can the perception of vegan leather be made more valuable and desirable for luxury consumers?

Identifying where the value lies goes back again, to some extent, to the definition of what constitutes a luxury product, premium, masstige, and fashion (described in previous chapters of this book), consumer motivations, and consumer groups (described earlier in this chapter).

References

Castle, P. (2022). *Louis Vuitton helping to polish the tarnished illegal exotic animal trade*. Retrieved March 10, 2024, from https://lampoonmagazine.com/article/2022/07/17/louis-vuitton-exotic-animal-trade/

Chandrasekhar, A. (2020). Retrieved March 15, 2024, from https://www.swissinfo.ch/eng/business/is-it-time-to-say-goodbye-to-exotic-leather-watch-straps/45920350

Dubois, B., & Laurent, G. (1994). Attitudes toward the concept of luxury: An exploratory analysis. *ACR Asia-Pacific Advances, 1*(2), 273–278.

Global Times. (2014). Retrieved March 13, 2024, from https://www.globaltimes.cn/content/887721.shtml

Hofstede, G. (1980). *Culture's consequences: International differences in work-related values*. Sage.

Kahn, R. (2011). *Towards an animal standpoint: Vegan education and the epistemology of ignorance. Epistemologies of ignorance and the studies of limits in education*. Information Age Publishing.

Kapferer, J. N. (2012). Abundant rarity: The key to luxury growth. *Business Horizons, 2012*(55), 453–462.

Keech, J., Morrin, M., & Podoshen, J. S. (2020). The effects of materialism on consumer evaluation of sustainable synthetic (lab-grown) products. *Journal of Consumer Marketing, 37*(5), 579–590. https://doi.org/10.1108/JCM-09-2018-2876

Kumar, S., & Gautam, A. (2021). Country of origin effect on purchase intention towards Italian luxury fashion: Mediating role of brand perception and social status. *Metamorphosis, 20*(1), 16–24. https://doi.org/10.1177/09726225211030067

Leather Working Group. (n.d.). Retrieved March 10, 2024, from https://www.leatherworking-group.com/

Leibenstein, H. (1950). Bandwagon, snob, and Veblen effects in the theory of consumers' demand. *Quarterly Journal of Economics, 64*(May), 183–207.

Lvmh. (2024). Retrieved March 10, 2024, from https://www.lvmh.com/houses/fashion-leather-goods/loro-piana/#:~:text=Based%20in%20northern%20Italy%2C%20Loro,its%20mission%20for%20six%20generations.

Michaut, A., Kapferer, J. N. (2020). *Sustainable Luxuries? Millennials are Skeptical—Yet They Buy*. https://www.hec.edu/en/sustainable-luxuries-millennials-are-skeptical-yet-they-buy

Par, L. C., & Nunes, M. F. (2024). Vegan luxury for non-vegan consumers: Impacts on brand trust and attitude towards the firm. *Journal of Retailing and Consumer Services, 77*(2024), 103669. https://doi.org/10.1016/j.jretconser.2023.103669

Piron, F. (2000). Consumers' perceptions of the country-of-origin effect on purchasing intentions of (in)conspicuous products. *Journal of Consumer Marketing, 17*(4), 308–321. https://doi.org/10.1108/07363760010335330

Ramchandani, M., & Coste-Manière, I. (2012). Asymmetry in multi-cultural luxury communication: A comparative analysis on luxury brand communication in India and China. *Journal of Global Fashion Marketing, 3*(2), 89–97. https://doi.org/10.1080/20932685.2012.10593111

Rogers, E. M. (1962). *Diffusion of innovations*. Free Press.

Vigneron, F., & Johnson, L. W. (2004). Measuring perceptions of brand luxury. *Journal of Brand Management, 11*(6), 484–506.

Vitro Labs Inc. (2024). Retrieved March 10, 2024, from https://www.vitrolabsinc.com/

Wiedmann, K. P., Hennigs, N., & Siebels, A. (2007). Measuring consumers' luxury value perception: A cross-cultural framework. *Academy of Marketing Science Review., 7*.

Chapter 7
Sustaining the Vegan Leather Industry: A Product-Type Assessment

1 Introduction

To use a vegan alternative for leather, it is necessary to determine what is expected of these materials in terms of applicability and performance. Physical, mechanical, and chemical properties determine the quality and utilization of materials, including traditional animal leather and vegan leather. In animal-based leather, different animal skins used and tanning procedures determine the quality and manufacturing of the finished products. Traditional animal-based leather, as described in previous chapters, is versatile and easy to repair, serving as a universal material for various applications. However, the environmental impact is highest for animal leather due to harmful pollutants, tanning processes, water wastage, and energy emissions, among other factors. Previous research (Ferreira da Silva, 2023) found that bio-based vegan alternatives like Piñatex (pineapple-based vegan leather), Desserto® (cactus-based vegan leather), and mycelium (fungi-based vegan leather) rank better in terms of environmental impact, such as life cycle assessments, but lag behind in durability compared to bovine leather, especially in the footwear industry. A study conducted (Spinnewijn, 2022) on innovative new vegan alternatives for leather discussed applications for different types of vegan leather in terms of mechanical strength and comparison with animal leather. The study found that Reishi™ (vegan leather from mycelium), Piñatex, and Desserto perform similarly to animal leather regarding technical properties and meets the requirement for shoe uppers and upholstery used in furniture, respectively. There are also differences between applications of different types of animal leather; for example, previous research (Ali et al., 2020) shows that cow suede leather is better than goat suede leather for making shoe uppers and linings due to better tensile strength, tearing strength, flexibility, and water vapor permeability. Despite the versatility of product applications and uses of

---

Mukta Ramchandani is the primary author of this chapter.

S. S. Muthu, M. Ramchandani, *Vegan Alternatives for Leather*,
Sustainable Textiles: Production, Processing, Manufacturing & Chemistry,
https://doi.org/10.1007/978-3-031-65365-0_7

different types of animal leather, they face several environmental challenges (as described in previous chapters). Therefore, newer sustainable vegan materials are being developed to cater to the high demand for leather in different industries and products that currently use leather.

2 How Different Vegan Leather Materials Are Used

For different types of vegan leathers as well, different requirements must be met. For example, to make bags, the material must be soft; for shoes, flexibility, abrasion resistance, adhesion resistance, and breathability of the material are expected. If each material is customized for different application during production, it will not only be sustainable but also lead to less wastage. There is significant wastage of animal leather when small products like bags are made (around 20%), and for the furniture and automotive industries, it ranges from 30% to 45%, respectively. This is because any natural marks on the animal hide are discarded, and only the parts that appear perfect are used by the brands. Consequently, this also affects the pricing of the product.

We gained insights on this topic from three interviews (detail methodology described in previous chapters).

In the case of Desserto, each material is made with a purpose, and the price comes from a lot of variables. For example, leather for yachts requires high UV and humidity resistance and salinity resistance. For bags, it needs to be softer and resistant to cosmetics (see Fig. 7.1). Each material is produced specifically to meet the specific requirements of customers. Additionally, customer expectations regarding sustainability, quality, and price are catered to, and a solution is presented to them that fulfills these needs comprehensively (Velarde, interview, 2023). See Figs. 7.2 and 7.3 for different colors and textures from Desserto.

Fig. 7.1 Bag made from Desserto cactus leather

Fig. 7.2 Desserto in several colors

In the case of Hanji paper used for making pouches by KHJ Studio in South Korea, two types of Hanji paper are used: gray and white color variations. To achieve a floral pattern, UV printing is used on the surface (see Figs. 7.4 and 7.5); the gray one has a natural stripped texture (see Figs. 7.6 and 7.7). Screen printing is used to emboss the logo in silver and gold. Further steps include cutting into shapes, sewing, and attaching sliders. The ideation, design, and development take about 6 months. The price varies with the quantity needed by clients. Some challenges associated with Hanji material include unsuitability for mass production; being handmade, it has a texture that is not conducive to machine printing, thus requiring traditional printing methods (Hyunjoo, interview, 2023).

In the case of the French footwear brand ÉMZI Paris, the company focuses on using natural materials (as detailed in previous chapters). The "Bambi Linen" shoe (see Fig. 7.8) uses linen sourced from a certified "Master of Linen" factory in Spain and is manufactured in Europe. The upper of this shoe is thick linen, while the heel and pointed-toe design are made of animal leather (REACH compliant and made in Europe) for protection, as linen alone could be too fragile. However, such footwear made of linen faces challenges, such as unsuitability for use in winter or rainy weather (Zhou Carpentier, interview, 2023).

Therefore, the construction of certain products may involve using specific parts to maintain functionality and performance. For example, in certain vegan shoes, a combination of natural and artificial vegan materials is used for the exterior upper and inner lining to ensure robustness and desired performance.

Fig. 7.3 Desserto's different texture

Fig. 7.4 KHJ Studio's pouch

3 Products Made from Vegan Leather for Different Industries

3.1 Automotive Industry

DuPont led the market with artificial vegan leather (Fabrikoid: cotton fabric coated with nitrocellulose) in collaboration with General Motors for car seats sold in the USA, especially in the early twentieth century (Mauddi, 2016). Most car brands have since used PU- and PVC-based materials (e.g., Naugahyde, MB-Tex) for

Fig. 7.5 KHJ Studio's floral pouches

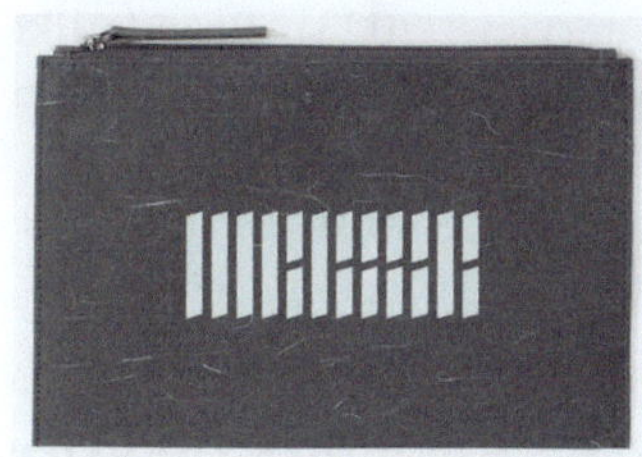

Fig. 7.6 KHJ Studio's gray pouch

upholstery. Currently, car brands are transitioning to natural/agro-based or bio-based vegan alternatives; for example, BMW uses Dessertex and Mirum.

3.2 Fashion Industry

Rubber, for example, can be used for lower parts, insoles, and outer soles of footwear (e.g., PLIANT™). Cork is also used in certain parts of shoes, bags, wallets, etc. Limited data is used for belts, as not all types of vegan leather are suitable. For instance, apple skin and fruit-based vegan leather do not last as long as animal leather, so belt manufacturers do not widely promote them (MIPEL, personal communication, 2022). Fleather, the flower-based leather, has limitations in maximum width and tensile strength required for belts (as described in Chap. 4). Adriano Di Marti, the company producing Desserto, can manufacture cactus leather material for belts with an industrial-level width of 1.4 meters (Velarde,

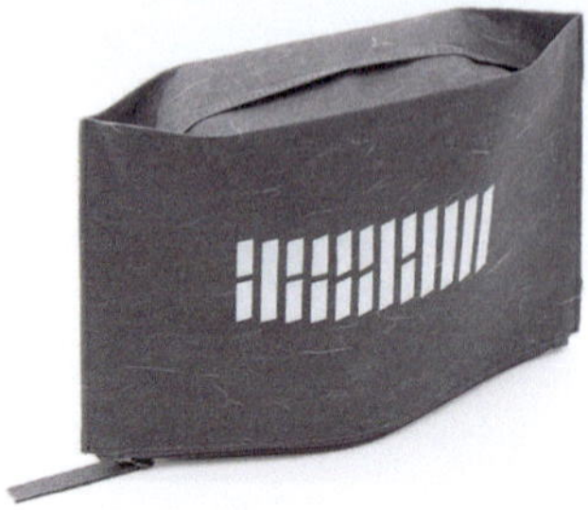

Fig. 7.7 KHJ Studio's gray pouch

Fig. 7.8 ÉMZI's Bambi Linen shoes

interview, 2023). Brands like Stella McCartney design vegan leather bags and accessories using the vegan leather alternative Mirum. Onitsuka Tiger used Desserto for the MEXICO 66 CACTFUL collection (Onitsuka Tiger, 2024), which was worn by Japanese athletes.

3.3 Furniture Industry

Various vegan leathers such as Desserto and Piñatex are used in the furniture industry. Limited data is available on other vegan alternatives. Artificial PU and PVC have long been used in interior and furniture designs.

3.4 Saddlery

For equestrians interested in vegan alternatives, there are currently limited options (to our knowledge), mainly with PU- and PVC-based faux leather saddlery options.

We also identified different products made from various vegan alternatives based on publicly available data from their websites, social media, and previous research publications (Table 7.1).

Table 7.1 Different products made from different vegan leather

Product categories	Desserto®/ Dessertex® cactus leather	Piñatex® pineapple leather	Hanji	Mirum®	Cork	Fungi-based mycelium (Reishi™)	PU/ PVC	Fleather flower-based leather
Bags	✓	✓	✓	✓	✓	✓	✓	✓
Wallets/ pouches	✓	✓	✓	✓	✓	✓	✓	✓
Belts	✓	No data	✗	No data	✓	No data	✓	✗
Shoes/ footwear	✓	✓	✗	✓	✓	✓	✓	No data
Clothes/ apparel	✓	✓	✗	✓	✓	✓	✓	No data
Automotive (interiors)	✓	No data	✗	✓	✗	No data	✓	No data
Furniture	✓	✓	✗	✓	✓	✓	✓	No data
Saddlery	No data	No data	✗	No data	✗	No data	✓	No data

4 Sustainability Attributes: Biodegradability, Longevity, Repairability, and Recyclability

Biodegradability is a controversial topic in material sustainability. Many argue that traditional leather is biodegradable and therefore sustainable. However, leather is only biodegradable in its natural condition; it does not completely biodegrade when processed with chemicals and tanning agents like chrome and chromium, which are carcinogenic. This is why there is emphasis from companies and governments (especially in Europe) on regulating tanning agents, restricting the use of certain chemicals in leather manufacturing, and adopting sustainable tanning methods (Guida et al., 2019). For sustainability in the supply chain, the German Supply Chain Due Diligence Act (Lieferkettensorgfaltspflichtengesetz (LkSG)) was recently passed in 2023. According to this act, companies in Germany with more than 3000 employees must be transparent and report on all human rights and environmental violations within their own operations and those of their direct supplier (BAFA, 2023).

While natural materials or materials containing a majority of natural ingredients are biodegradable as organic matter decomposes when exposed to UV, humidity, enzymatic activities of microorganisms, and other conditions, the biodegradability factor also varies from product to product. For instance, leather used in automotive industry interiors is biodegradable under specific conditions but must also meet quality requirements for daily use without failing. SCOBY-based vegan materials derived from kombucha tea are naturally biodegradable, but when processed to make them water-resistant with the application of oils and acrylics, the materials lose their biodegradable properties (Payne et al., 2016).

Material biodegradability is also crucial in the final phases of a product's life cycle, whether in the elimination phase or the recyclability phase. From a product circularity perspective, the recyclability of a material enhances its sustainability attributes. However, from the perspective of closing the loop in circularity, biodegradability also contributes to sustainability attributes. Moreover, as described in previous chapters, each vegan material differs in its applicability; therefore, these aspects of biodegradability and recyclability must be balanced.

Since different vegan leather alternatives are formulated differently, each requires different recycling standards. Companies or brands using vegan leather materials can refer to third-party certifications like the RCS (Recycled Claim Standard) and GRS (Global Recycled Standard) to verify and assess the recyclability of materials both pre- and post-consumption. Often, the recyclability also depends on how the final product is used by the customer and constructed. For instance, if a product using vegan leather is lined with synthetic material or uses certain unsustainable adhesives, recycling becomes challenging.

For Desserto and Dessertex, material sustainability and performance are primary focuses for Adriano Di Marti. These materials are biodegradable under specific conditions, allowing for safe decomposition or recyclability. Different compositions range from 100% bio-based to 30–90% plant-based materials, such as organic cotton lining in some products. According to Mr. Velarde (interview, 2023), products made from Desserto have received positive feedback. The materials are tested in the laboratory of UV resistance, abrasion, and humidity resistance, allowing their behavior to be studied for years. These materials comply with the most rigorous requirements, not only for leather but also for synthetics. It is crucial to inform end customers about proper care to ensure longevity. Just as corrosive agents can harm skin, they can damage any material, including leather. Desserto materials are easy to clean only with just a cloth or a bit of water, ensuring durability.

Velarde further stated that while there is much discussion about biodegradability, product performance is often overlooked. The priority is to create materials that are as organic as possible while meeting quality standards. Engineering work is then done to make them biodegrade under specific conditions, allowing the material to decompose in a safer way or to be recycled. Most of their research focuses on understanding the biodegradability behavior of their materials in relation to their plant-based content, ensuring that they do not fail in daily use.

For repairability, certain vegan materials are repairable but to an extent with certain complexities. For example, Hanji paper pouches can be repaired, but the material starts aging and looks old over the years (Hyunjoo, interview, 2023).

To repair Desserto, it is easy to cut and sew and recycling is also possible at the industrial level. Transforming it into an intermediary raw material for producing new materials is also feasible but requires a sophisticated process (Velarde, interview, 2023).

We also assessed the sustainability attributes of different vegan alternatives based on publicly available data from their websites, social media, and previous research publications (Table 7.2).

Table 7.2 Sustainability attributes of different vegan leather

Sustainability attributes	Desserto®/ Dessertex® cactus leather	Piñatex pineapple leather	Hanji	Mirum	Cork	Fungi-based mycelium (Reishi™)	PU/PVC	Fleather flower-based leather
Biodegradability	Partially under controlled conditions	Partially	✓	Partially	✓	✓	✘	✓
Durability (short/ moderate/long-term)	Moderate-long	Moderate	Short	Moderate-long (limited data)	Short	Moderate	Moderate	Short
Repairability	✓	Partially repairable	Partially repairable	No data	✘	No data	✘	✘
Recyclability	✓	Moderate	Moderate	No data	✓	No data	✘	✘
Life cycle assessment	✓	✓	No data	✓	✓	✓	✓	✘

5 Discussion and Conclusion

From discussions with several managers and industrial data, the perception around vegan leather has been that it is not better than animal leather in terms of performance. Many designers who adhere to sustainable standards prefer using natural, animal-based leather options that are eco-friendly, such as vegetable-tanned and REACH compliant, over vegan leather. Some brands, use both animal leather and vegan leather (e.g.: Veja). Several researchers (Minh & Ngan, 2021; Meyer et al., 2021; Saha et al., 2020) also point out that while naturally derived vegan leather lags behind in performance compared to artificial faux leather (PU and PVC) and animal-based leather, it still considered more sustainable, especially with LCA data. Material performance encompasses the technical aspects of a material, including its mechanical and chemical properties. Mechanical properties include thickness, tensile strength, tear strength, flex resistance, water vapor permeability, and absorption (Meyer et al., 2021). Chemical properties involve the chemical composition of the material, cosmetic resistance, UV resistance, humidity resistance, etc. Therefore, maintaining a balance between material performance and sustainability is crucial (see Fig. 7.9 for a summary).

The debate in the industry revolves around the balance between material performance and the sustainability of vegan leather. While a vegan material may be entirely eco-friendly and sustainable, it may not perform well for certain products (e.g., Fleather or Hanji, which are not strong enough for making shoes, clothes, belts, or furniture) and are produced in limited quantities. On the other hand, vegan materials like PU and PVC can be used for a variety of applications but are among the most unsustainable and environmentally harmful materials. Moderately sustainable options with a moderate number of product applications and significant performance include natural/agro-derived or mycelium-based vegan leathers (see Table 7.2).

Therefore, based on previous literature and performance assessments, animal leather remains a suitable choice (see Fig. 7.10 for a ranking of sustainability and

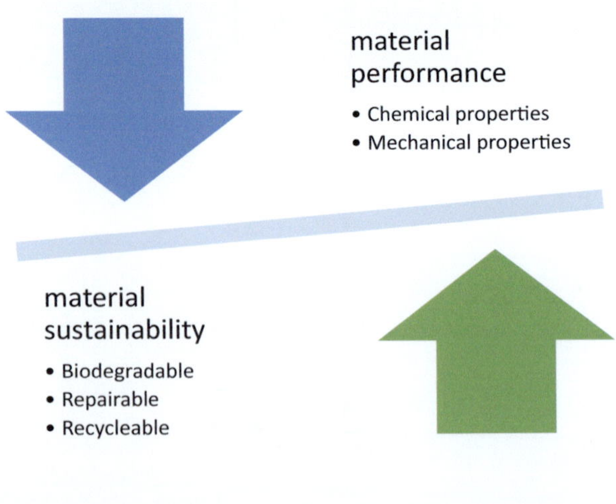

Fig. 7.9 Balance between material performance and sustainability

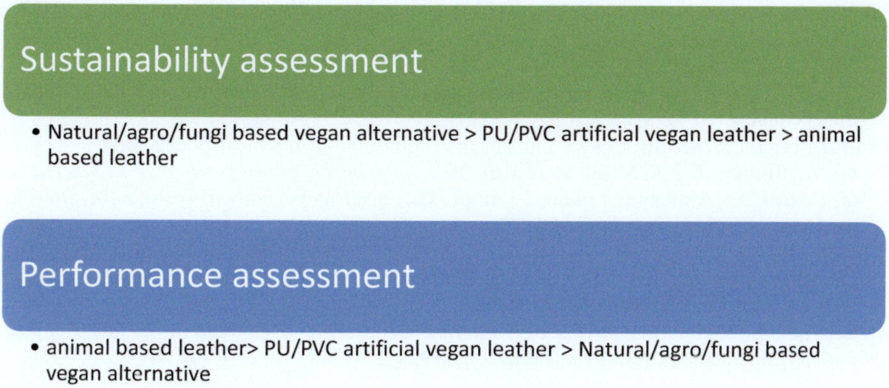

Fig. 7.10 Ranking of sustainability and performance assessment

performance assessments). However, considering overall sustainability assessments that integrate animal welfare perspectives and environmental impacts (such as pollution, water treatment, waste, carbon footprint, LCA data, biodegradability, repairability, and recyclability), some sustainable vegan leather materials like natural/agro-based and bio-based vegan leathers (Dessertex, Mirum, Reishi, Cork, Fleather, Hanji) are better choices than animal leather and artificial PU/PVC leather. There is one exception to this sustainability assessment criterion: "longevity," as some vegan materials are relatively new to the market and limited research data is available on their long-term durability and ability to be passed down through generations, unlike some luxury animal leather products (explained in previous chapters).

References

Ali, F., Kamal, M., & Islam, M. S. (2020). Comparative study on physical properties of different types of leather in Bangladesh. *International Journal of Engineering Research and Applications, 10*(2), 55–63. https://doi.org/10.9790/9622-1002035563

BAFA. (2023). Retrieved March 4, 2024, from https://www.bafa.de/EN/Supply_Chain_Act/supply_chain_act_node.html

Ferreira da Silva, P. P. (2023). *Environmental performance of leather and leather-like materials applied to footwear*. NTNU. https://ntnuopen.ntnu.no/ntnu-xmlui/handle/11250/3087821

Guida, M., Giorgio, A., Aliberti, F., Libralato, G., & De Bonis, S. (2019). Biodegradability of eco-friendly leather using respirometric approach. *Journal of the American Leather Chemists Association, 114*(9), 325–332.

Mauddi, N. (2016). *DuPont and the Rise of Faux Leather*. https://www.thehogring.com/2016/12/21/dupont-and-the-rise-of-faux-leather/

Meyer, M., Dietrich, S., Schulz, H., & Mondschein, A. (2021). Comparison of the technical performance of leather, artificial leather, and trendy alternatives. *Coatings, 11*, 226. https://doi.org/10.3390/coatings11020226

Minh, N. T., & Ngan, H. N. (2021). Vegan leather: An eco-friendly material for sustainable fashion towards environmental awareness. *AIP Conference Proceedings, 2406*(1), 060019. https://doi.org/10.1063/5.0066483

MIPEL. (2022). *Belts manufacturer*, Personal communication, MIPEL Milano Italy, 2022. https://mipel.com/en/

Onitsuka Tiger. (2024). https://www.onitsukatiger.com/us/en-us/mk/mexico66cactful#

Payne, A, Brough, D., & Musk, P. (2016). *Will we soon be growing our own vegan leather at home?* Available online: https://theconversation.com/will-we-soon-be-growing-our-own-vegan-leather-at-home-68498

Saha, N., Ngwabebhoh, F. A., Nguyen, H. T., & Saha, P. (2020). Environmentally friendly and animal free leather: Fabrication and characterization. *AIP Conference Proceedings, 2289*(1), 020049. https://doi.org/10.1063/5.0028467

Spinnewijn, O. (2022). *Innovative leather alternatives: A short- lived trend or the future of fashion?* https://studenttheses.uu.nl/handle/20.500.12932/41443

Chapter 8
Conclusion and Outlook

1 Role in Shaping Up the Industry

Vegan alternatives for leather have been assessed and discussed in the previous chapters of this book. We have assessed the role vegan alternatives play in terms of types of different vegan leathers, sustainability, production processes, usability, consumer behavior, and perception from the luxury and fashion sectors. We have shown that it is not just about sustainability but rather finding a balance between sustainability and performance that these materials need to be assessed.

1.1 What Is the Physical Advantage of Vegan Leather Compared to Animal Leather? Is It Sustainable?

Vegan leather presents a compelling alternative to animal leather, boasting several physical advantages and sustainability benefits, although it does encounter some challenges.

1.1.1 Physical Advantages

- *Uniformity:* Unlike animal leather, vegan leather offers a consistent appearance, facilitating more predictable manufacturing processes (Softline, n.d.).
- *Variety of textures and colors:* With a diverse range of textures and colors, vegan leather caters to various design preferences (Ashraf, 2023).

S. S. Muthu, M. Ramchandani, *Vegan Alternatives for Leather*,
Sustainable Textiles: Production, Processing, Manufacturing & Chemistry,
https://doi.org/10.1007/978-3-031-65365-0_8

- *Ease of maintenance:* Certain types of vegan leather, especially those not derived from PVC, are easier to clean and maintain compared to animal leather (Softline, n.d.).
- *Lightweight:* Vegan leather tends to be lighter, enhancing comfort and wearability in fashion applications (Mahi Leather, n.d.)

1.1.2 Sustainability

Advantages

- *Reduced environmental mpact:* Production methods for vegan leather, particularly those utilizing plant-based sources, often result in lower environmental footprints compared to animal leather (Davis, 2020).
- *No animal cruelty:* Vegan leather eliminates concerns related to animal welfare (Ashraf, 2023).
- *Innovation in materials:* Ongoing advancements, such as pineapple leaf-based Piñatex and mushroom mycelium leather, contribute to sustainability and reduce reliance on fossil fuels (Ramos, 2023).

1.1.3 Challenges

- *Dependence on fossil fuels:* Some vegan leather variants, notably those containing PVC and PU, rely on fossil fuel-derived materials, raising sustainability questions (Mayhead, n.d.).
- *Biodegradability:* Synthetic vegan leathers may not readily decompose and could release harmful chemicals during breakdown (Pangilinan, 2023).
- *Durability concerns:* While high-quality vegan leather can be durable, it may not match the longevity of animal leather, potentially leading to more frequent replacements and waste generation (Softline, n.d.).

In summary, vegan leather and traditional leather have their own set of advantages and disadvantages. Both can coexist harmoniously in the market, catering to diverse customer preferences for a varied range of products.

According to the research by Meyer et al. (2021), none of the vegan leather materials they analyzed (e.g., Desserto®, Kombucha, Piñatex®, Noani®, AppleSkin®, VEGEA®, SnapPap®, Teak Leaf®, and MuSkin®) pose any threat to genuine leather. They assessed them based on structural composition, technical performance, mechanical properties, water vapor permeability, and water absorption. From our understanding while writing this book and researching this topic, as well as our previous experience in the sector, none of the vegan leather materials are better in performance than traditional leather. So what is the solution for the industry? Our proposition for the industry is that it will still be better to use some of the natural/agro-based and fungi-based vegan leather materials where appropriate in application. We do not advocate using PU/PVC vegan leather, which has been found through extensive research to be extremely damaging to the environment (see previous chapters). Also, we do not advocate using fruit-based vegan leather like mango/apple/grape as these are neither sustainable nor free from pesticides and are in direct conflict with the food industry and farming. Following are some of the recommendations for different industries:

- For *mass products in fashion*, which anyway depend on cheaper prices, faster consumption cycles, and following trends (see Chaps. 4 and 6), vegan alternatives with *faster to moderate biodegradability* will be best (e.g., for items like wallets, Fleather (flower-based leather), Hanji paper, cork; for bags and shoes, Desserto and Piñatex). The specific attributes of these materials were discussed in previous chapters (Chaps. 6 and 7).
- For the *luxury industry*, because the attributes of a luxury product are very specific—rarity, craftsmanship, quality, exclusivity, longevity, timelessness, and high prices—it must be differentiated not only based on sustainability attributes but also by consumer group: vegan or non-vegan (see Chap. 6).

For vegan consumers, based on the currently available vegan alternatives which are also used by some luxury brands like Stella McCartney and Hermès, it is limited to Mirum® and some mycelium-based materials (e.g., Reishi™ and Mylo). For other vegan alternatives, there is still very limited data, and more research is needed to forecast the long-term usage and luxury appeal of the materials.

For non-vegan consumers, options are animal-based leather vegetable-tanned, which is sustainably sourced and limitedly produced, or the new lab-made animal leather bioengineered without killing animals but using stem cells (e.g., VitroLabs Inc). However, limited research data is available for this new type of lab-made animal leather.

- For the *automotive industry*, the emphasis is on longer usage with high sturdy materials and strong mechanical properties from the currently available vegan materials, Dessertex® and Mirum®.
- For the *furniture industry*: cork, Desserto, Piñatex, Mirum.

2 Toxicity of Chemicals

As discussed in Chaps. 2 and 3 of the book, the hazardous chemicals used in traditional leather tanning (chromium, formaldehyde, azo dyes, sodium sulfide, chlorinated paraffin, heavy metals, perfluorinated compounds, and solvents) not only pollute the environment but also cause serious health issues like cancer for workers.

The use of chromium salts in traditional leather tanning, predominantly chromium (III), poses potential risks. When oxidized, chromium (III) can transform into chromium (VI), a recognized toxin and carcinogen. Chromium (VI) exposure is associated with severe health complications such as skin ailments, ulcers, and heightened lung cancer risk (Damini, n.d.).

This does not mean that in the vegan leather industry, none of the chemicals mentioned above are used (refer to detailed LCA comparison data as shown in Chap. 3).

There is a European regulation that bans the usage of AZO dyes, and leather must adhere to REACH compliance. For the vegan alternatives as well, these need to be followed. In one of the interviews that we collected for this book, we asked this question to the co-founder of Adriano Di Marti producing Desserto and Dessertex. According to Mr. Velarde, they use organic pigment instead of a dyeing process. While incorporating pigments, it must be ensured that they do not contain any heavy metals, have quality and resistance for UV color fastness, and contain nothing from animals. When selling materials to brands, they ask them to use water-based adhesives because using solvents can be harmful to the material and cause corrosion. Not every aspect can be controlled after the materials are sold to other companies.

3 A Threat to Traditional Leather? What Is the Future?

The newer vegan materials are made with different technologies and have found their applications in different sectors, which also shows that the plastic-based image of vegan leather is changing and is only restricted to PU- and PVC-based vegan materials. For brands and designers looking to adopt vegan leather, several challenges and how to mitigate those challenges are discussed in Chap. 5. Importantly, skillsets need to be transferred from different regions to make the vegan leather industry more commercially viable.

Consumer types and perceptions must be further researched regarding sustainable vegan leather options from both academia and industry. More communication and transparency are needed from vegan leather brands and retailers, especially regarding how the vegan leather was made, how long the materials last, and information about the sustainability attributes of vegan leather products. Just marketing a vegan leather product as "cruelty-free" is not enough for all consumers; sustainability plays a larger and holistic role in protecting animals, the planet, and its people.

From the data we assessed and the three interviews we collected, the conclusion we reached is that traditional animal-based leather is very versatile in applications and performance and has high demand, but at the same time, it has serious environmental issues, especially due to its mass production and fast-fashion type of usage. Vegan alternatives for leather are here to stay; currently, not all are scalable to a level that meets the demand of the leather industry, but they undeniably push the leather industry to be more sustainable and innovative.

The future of vegan leather looks promising, with potential for further growth and innovation. As technology advances, the quality and texture of vegan leather are expected to improve, making it an even more attractive alternative to traditional leather. The industry's growth is also likely to be supported by regulatory support and changing consumer preferences toward more sustainable and cruelty-free products (Vantage Market, n.d.).

Vegan leather is not poised as a threat to genuine leather; rather, it caters to a distinct market segment driven by values such as sustainability, ethics, and cruelty-free production. Its ascent mirrors a consumer shift toward these principles, spurred by concerns over animal welfare and environmental impact. Ongoing research and development are enhancing the quality and sustainability of vegan leather, with innovations like mushroom and cactus leather opening new market avenues. However, challenges persist, including consumer skepticism regarding quality. Despite its rise, traditional leather retains its allure due to attributes like durability and aesthetics. Moreover, the leather industry is adapting, integrating more sustainable practices, and exploring eco-friendly tanning methods to address environmental concerns (Urban Southern, 2018). There should be many product-level assessments based on LCA and social and environmental supply chain assessments for these leather alternatives to bring out the potential environmental benefits of such alternatives vis-a vis conventional leather.

In summary, vegan leather stands at the forefront of a transformative movement toward sustainability and ethics in fashion and beyond (VOES & Co, n.d.). Its growing significance mirrors a collective shift among consumers, designers, and industries toward environmentally conscious and animal-friendly alternatives. Embracing and advancing vegan leather technology is imperative for diminishing the environmental impact of fashion and steering us toward a more sustainable path forward.

References

Ashraf, R. (2023). *Blackbird leathers—Real leather vs. Vegan leather | A deep analysis*, Retrieved April 14 2024, from https://blackbirdleathers.com/fashion-and-compassion-real-leather-vs-vegan-leather/

Damini. (n.d.). *Deskera—Leather chemicals and their impact on the environment*. Retrieved April 13, 2024, from https://www.deskera.com/blog/leather-chemicals-and-their-impact-on-the-environment/

Davis, J. (2020). *Is Vegan leather worse for the environment than real leather?* Retrieved April 13, 2024, from https://www.harpersbazaar.com/uk/fashion/fashion-news/a30640996/vegan-leather-sustainability/

Mahi Leather, (n.d.). *All you need to know about Vegan Leather | What is it? How is it made?—MAHI leather*, Retrieved April 13, 2024, from https://mahileather.com/blogs/news/all-you-need-to-know-about-vegan-leather,

Mayhead, J. (n.d.). *Ethical leather—What is Vegan leather, and is Vegan leather sustainable?* Retrieved April 13, 2024, from https://ethicalmadeeasy.com/is-vegan-leather-actually-better-for-the-environment/

Meyer, M., Dietrich, S., Schulz, H., & Mondschein, A. (2021). Comparison of the technical performance of leather, artificial leather, and trendy alternatives. *Coatings, 11*, 226. https://doi.org/10.3390/coatings11020226

Pangilinan, K. (2023). *Kind pet—Is Vegan leather really sustainable?* Retrieved April 13, 2024, from https://thekindpet.com/blogs/blog/is-vegan-leather-really-sustainable#:~:text=Vegan%20leather%20is%20all%20about,bit%20of%20an%20environmental%20hiccup

Ramos, D. (2023). *Analysing the Pros and Cons of Vegan leather*. Retrieved April 13, 2024, from https://earth.org/analysing-the-pros-and-cons-of-vegan-leather/.

Softline. (n.d.). *Soft line Vegan leather and its benefits*. Retrieved April 13, 2024, from https://softlinebrandpartners.com/what-is-vegan-leather-and-what-are-its-benefits-over-traditional-leather/#:~:text=Plant%2Dbased%20leather%20can%20be,cutting%20yield%20far%20more%20predictable.

Urban Southern. (2018). *3 key differences between authentic leather and vegan leather.* Urban Southern, Retrieved April 13, 2024, from https://urbansouthern.com/blogs/urban/differences-authentic-leather-vegan-leather/

Vantage Market. (n.d.). *Vegan leather market size USD 131.7 Million by 2030*, Retrieved April 13, 2024, from https://www.vantagemarketresearch.com/industry-report/vegan-leather-market-2052#:~:text=Snapshot&text=The%20global%20Vegan%20Leather%20Market,is%20often%20produced%20from%20polyurethane.

Voes & Co. (n.d.). *The rise of Vegan leather: A fashion revolution—VOES & CO*. Retrieved April 13, 2024, from https://voesandcompany.com/blogs/news/the-rise-of-vegan-leather-a-fashion-revolution

Index

S. S. Muthu, M. Ramchandani, *Vegan Alternatives for Leather*, Sustainable Textiles: Production, Processing, Manufacturing & Chemistry,
https://doi.org/10.1007/978-3-031-65365-0